166/2019

MEMOIRS
of the
American Mathematical Society

ISSN 0065-9266 (print)
ISSN 1947-6221 (online)

Number 1234

0 000 000 143232 F

AF848529

Interpolation for Normal Bundles of General Curves

Atanas Atanasov
Eric Larson
David Yang

The Abdus Salam International Centre for Theoretical Physics
LIBRARY
14 JUN 2019
Strada Costiera, 11
Trieste ITALY

January 2019 • Volume 257 • Number 1234 (fourth of 6 numbers)

AMERICAN MATHEMATICAL SOCIETY

Library of Congress Cataloging-in-Publication Data

Names: Atanasov, Atanas, 1988– author. | Larson, Eric, 1991– author. | Yang, David, 1997– author.

Title: Interpolation for normal bundles of general curves / Atanas Atanasov, Eric Larson, David Yang.

Description: Providence, RI : American Mathematical Society, [2018] | Series: Memoirs of the American Mathematical Society, ISSN 0065-9266 ; volume 257, number 1234 | Includes bibliographical references.

Identifiers: LCCN 2018053269 | ISBN 9781470434892 (alk. paper)

Subjects: LCSH: Curves, Algebraic. | Projective spaces. | Geometry, Projective.

Classification: LCC QA565 .A83 2018 | DDC 516.3/52–dc23

LC record available at https://lccn.loc.gov/2018053269

DOI: https://doi.org/10.1090/memo/1234

Memoirs of the American Mathematical Society

This journal is devoted entirely to research in pure and applied mathematics.

Subscription information. Beginning with the January 2010 issue, *Memoirs* is accessible from `www.ams.org/journals`. The 2019 subscription begins with volume 257 and consists of six mailings, each containing one or more numbers. Subscription prices for 2019 are as follows: for paper delivery, US$1041 list, US$832.80 institutional member; for electronic delivery, US$916 list, US$732.80 institutional member. Upon request, subscribers to paper delivery of this journal are also entitled to receive electronic delivery. If ordering the paper version, add US$20 for delivery within the United States; US$80 for outside the United States. Subscription renewals are subject to late fees. See `www.ams.org/help-faq` for more journal subscription information. Each number may be ordered separately; *please specify number* when ordering an individual number.

Back number information. For back issues see `www.ams.org/backvols`.

Subscriptions and orders should be addressed to the American Mathematical Society, P. O. Box 845904, Boston, MA 02284-5904 USA. *All orders must be accompanied by payment.* Other correspondence should be addressed to 201 Charles Street, Providence, RI 02904-2213 USA.

Copying and reprinting. Individual readers of this publication, and nonprofit libraries acting for them, are permitted to make fair use of the material, such as to copy select pages for use in teaching or research. Permission is granted to quote brief passages from this publication in reviews, provided the customary acknowledgment of the source is given.

Republication, systematic copying, or multiple reproduction of any material in this publication is permitted only under license from the American Mathematical Society. Requests for permission to reuse portions of AMS publication content are handled by the Copyright Clearance Center. For more information, please visit `www.ams.org/publications/pubpermissions`.

Send requests for translation rights and licensed reprints to `reprint-permission@ams.org`.

Excluded from these provisions is material for which the author holds copyright. In such cases, requests for permission to reuse or reprint material should be addressed directly to the author(s). Copyright ownership is indicated on the copyright page, or on the lower right-hand corner of the first page of each article within proceedings volumes.

Memoirs of the American Mathematical Society (ISSN 0065-9266 (print); 1947-6221 (online)) is published bimonthly (each volume consisting usually of more than one number) by the American Mathematical Society at 201 Charles Street, Providence, RI 02904-2213 USA. Periodicals postage paid at Providence, RI. Postmaster: Send address changes to Memoirs, American Mathematical Society, 201 Charles Street, Providence, RI 02904-2213 USA.

©2019 Atanas Atanasov, Eric Larson, and David Yang. All Rights Reserved.

This publication is indexed in *Mathematical Reviews®*, *Zentralblatt MATH*, *Science Citation Index®*, *Science Citation Index™-Expanded*, *ISI Alerting Services℠*, *SciSearch®*, *Research Alert®*, *CompuMath Citation Index®*, *Current Contents®/Physical, Chemical & Earth Sciences*.

This publication is archived in *Portico* and *CLOCKSS*.

Printed in the United States of America.

∞ The paper used in this book is acid-free and falls within the guidelines established to ensure permanence and durability.

Visit the AMS home page at `https://www.ams.org/`

10 9 8 7 6 5 4 3 2 1 24 23 22 21 20 19

Contents

Abstract

Given n general points $p_1, p_2, \ldots, p_n \in \mathbb{P}^r$, it is natural to ask when there exists a curve $C \subset \mathbb{P}^r$, of degree d and genus g, passing through $p_1, p_2, \ldots, p_n$. In this paper, we give a complete answer to this question for curves C with nonspecial hyperplane section. This result is a consequence of our main theorem, which states that the normal bundle N_C of a general nonspecial curve of degree d and genus g in $\mathbb{P}^r$ (with $d \geq g+r$) has the property of *interpolation* (i.e. that for a general effective divisor D of any degree on C, either $H^0(N_C(-D)) = 0$ or $H^1(N_C(-D)) = 0$), with exactly three exceptions.

Received by the editor November 4, 2015 and, in revised form, May 17, 2016; June 10, 2016; and June 11, 2016.

Article electronically published on January 9, 2019.

DOI: `https://doi.org/10.1090/memo/1234`

2010 *Mathematics Subject Classification.* Primary 14H99.

©2019 Atanas Atanasov, Eric Larson, and David Yang

CHAPTER 1

Introduction

The study of curves in projective space is one of the major topics in modern algebraic geometry. It has also served as a central example in the broader interest in moduli spaces which has flourished during the past half-century. The goal of the present article is to address the following fundamental question about incidence conditions for curves.

MAIN QUESTION. When is there a (smooth) curve of degree d and genus g passing through n general points in $\mathbb{P}^r$?

Several cases of this question, and of closely related questions we shall discuss below, have been previously studied in the literature. For example, the case of rational curves ($g = 0$) was answered independently by both Sacchiero [**10**] and Ran [**9**], and partial results for space curves ($r = 3$) were obtained independently by both Perrin [**8**] and Atanasov [**1**].

There are also several generalizations worth mentioning. For example, given values d, g, r, and n, we can ask for the dimension of the space of appropriate curves which satisfy the incidence conditions for a general collection of n points. Alternatively, we can also replace the points with with higher dimensional linear spaces, or even other subvarieties in projective space. It turns out that the main question and its generalizations are all related to a property of vector bundles over curves we call *interpolation*. If the normal bundle of a curve satisfies interpolation, we deduce a statement about the deformation theory of the curve, which in turn can lead to an answer of the main question.

Before going any further, we will elaborate the connection between our main question and interpolation of normal bundles. Our references are [**11**] and [**8**]. Let $\mathcal{H}_{d,g,r}$ and $\mathcal{P}_{n,r}$ respectively denote the Hilbert schemes of curves of degree d and genus g in $\mathbb{P}^r$, and n points in $\mathbb{P}^r$. There is an incidence correspondence $\Sigma \subset \mathcal{P}_{n,r} \times \mathcal{H}_{d,g,r}$ (a flag Hilbert scheme) whose points are pairs $([D], [C])$ such that $D \subset C$.

$$\begin{array}{ccc} & \Sigma & \\ {\scriptstyle f}\swarrow & & \searrow{\scriptstyle g} \\ \mathcal{P}_{n,r} & & \mathcal{H}_{d,g,r} \end{array}$$

Choose a point $([D], [C])$ such that C is an lci curve and $D \subset C$ is a Cartier divisor. There is an identification of tangent spaces $T_{[C]}\mathcal{H}_{d,g,r} \cong \mathrm{H}^0(N_C)$ and similarly for

D. Then the tangent space $T = T_{([D],[C])}\Sigma$ fits in the following Cartesian diagram.

$$\begin{array}{ccc} T & \xrightarrow{dg} & \mathrm{H}^0(N_C) \\ {\scriptstyle df}\downarrow & & \downarrow \\ \mathrm{H}^0(N_D) & \longrightarrow & \mathrm{H}^0(N_C|_D) \end{array}$$

THEOREM 1.1 (Kleppe). *Let $([D],[C])$ be a geometric point of Σ. If $[C] \in \mathcal{H}_{d,g,r}$ is a smooth point, and the restriction morphism $\mathrm{H}^0(N_C) \to \mathrm{H}^0(N_C|_D)$ is surjective, then f is smooth at the point $([D],[C])$. In particular, the image of f contains an open neighborhood of $[D]$.*

If the hypotheses of Theorem 1.1 are satisfied, then we can give a positive answer to the main question. Consider the short exact sequence

$$0 \longrightarrow N_C(-D) \longrightarrow N_C \longrightarrow N_C|_D \longrightarrow 0,$$

whose cohomology sequence reads

$$0 \longrightarrow \mathrm{H}^0(N_C(-D)) \longrightarrow \mathrm{H}^0(N_C) \longrightarrow$$

$$\mathrm{H}^0(N_C|_D) \longrightarrow \mathrm{H}^1(N_C(-D)) \longrightarrow \mathrm{H}^1(N_C) \longrightarrow 0.$$

If $\mathrm{H}^1(N_C(-D)) = 0$, then $\mathrm{H}^0(N_C) \to \mathrm{H}^0(N_C|_D)$ is surjective and $[C]$ is a smooth point of $\mathcal{H}_{d,g,r}$ (because $\mathrm{H}^1(N_C) = 0$), so we can apply Theorem 1.1. Note that if N_C is nonspecial, then $\mathrm{H}^1(N_C(-D)) = 0$ is equivalent to

$$\mathrm{h}^0(N_C(-D)) = \mathrm{h}^0(N_C) - (r-1)\deg(D). \tag{1.1}$$

This discussion naturally leads us to the definition of interpolation (see Theorem 4.1). In this particular case, the bundle N_C satisfies interpolation if

(1) for all $n \leq \mathrm{h}^0(N_C)/(r-1)$, there exists a degree n divisor D satisfying Eq. (1.1), and
(2) for all $n > \mathrm{h}^0(N_C)/(r-1)$, there exists a degree n divisor D such that $\mathrm{h}^0(N_C(-D)) = 0$.

Given a general curve C of genus g, and a general line bundle $\mathcal{L}$ on C of degree d, it is well-known that there exists a linear series on C attached to $\mathcal{L}$ defining a map to $\mathbb{P}^r$ if and only if

$$d \geq g + r.$$

Moreover, in this range, there is a unique component of the Hilbert scheme corresponding to such curves; this component is distinguished by the fact that a general curve in this component has a nonspecial hyperplane section (which we will refer to as a "nonspecial curve" for brevity). Our main result determines when the normal bundle of a general nonspecial curve satisfies interpolation:

THEOREM 1.2. *Let C be a general nonspecial curve of degree d and genus g in $\mathbb{P}^r$ (where $d \geq g + r$). Then the normal bundle N_C satisfies interpolation, unless:*

$$(d, g, r) \in \{(5, 2, 3), (6, 2, 4), (7, 2, 5)\}.$$

The condition of interpolation is equivalent for rational curves (and analogous in some sense for curves of higher genus) to the conditions of semistability and section-semistability (see Section 3 of [**1**]), although we shall not make use of these analogies here. However, we will remark that the analog of Theorem 1.2 for semistability of the normal bundle is known in the case of rational curves ($g = 0$) as mentioned earlier [**9, 10**], as well as in the case of linearly normal elliptic curves ($g = 1$ and $d = r + 1$) by work of Ein and Lazarsfeld [**4**].

As a consequence, we answer the main question posed at the beginning of the introduction for nonspecial curves:

COROLLARY 1.3. *There exists a nonspecial curve C of degree d and genus g in $\mathbb{P}^r$ (with $d \geq g + r$), passing through n general points, if and only if*

$$\begin{cases} (r-1)n \leq (r+1)d - (r-3)(g-1) & \text{if } (d, g, r) \notin \{(5, 2, 3), (7, 2, 5)\}; \\ n \leq 9 & \text{if } (d, g, r) \in \{(5, 2, 3), (7, 2, 5)\}. \end{cases}$$

To prove Theorem 1.2, we will argue by inductively degenerating C to a reducible nodal curve $X \cup Y$. We use results of Hartshorne and Hirschowitz [**6**] to guarentee the existance of particular such degenerations, and to give descriptions of the restrictions $N_{X \cup Y}|_X$ and $N_{X \cup Y}|_Y$. However, in order to reduce interpolation for N_C to statements about $N_{X \cup Y}|_X$ and $N_{X \cup Y}|_Y$, we need to have a geometric description of the gluing data:

$$H^0(N_{X \cup Y}|_X) \to H^0(N_{X \cup Y}|_{X \cap Y}) \leftarrow H^0(N_{X \cup Y}|_Y). \tag{1.2}$$

The key observation that makes it possible to approach Theorem 1.2 is the existence — in the case when $Y = L$ is a line — of certain geometrically-defined line subbundles $\mathcal{L} \subseteq N_{X \cup Y}$, which taken together enable us to give an essentially-complete geometric description of the gluing data in Eq. (1.2).

For example, suppose that $Y = L$ meets X in a single point u; write $v \in L$ for some point on L distinct from u. Then writing $S = \overline{v \cdot X}$ for the cone over X with vertex v, the normal bundle $\mathcal{L}$ of $X \cup L$ in S gives such a bundle. We will see in Chapter 8 that $\mathcal{L}|_L$ gives the positive subbundle of $N_{X \cup L}|_L$; using this, we will reduce interpolation for $N_{X \cup Y}$ to interpolation for the vector bundle on X given by the kernel of the natural map

$$N_{X \cup Y}|_X \to (N_{X \cup Y}|_X / \mathcal{L})|_u. \tag{1.3}$$

Summary. We begin the paper in Chapter 2 and Chapter 3 by studying *modifications* of vector bundles, which are generalizations of the above bundle defined on X — where $N_{X \cup Y}|_X$ is replaced an arbitrary vector bundle on X, and $\mathcal{L}$ by an arbitrary subbundle of $N_{X \cup Y}|_X$. That is, the modification $\mathcal{E}[D \to \mathcal{F}]$ of $\mathcal{E}$ along $\mathcal{F}$ at a Cartier divisor D is simply the kernel of the natural map

$$\mathcal{E} \to (\mathcal{E}/\mathcal{F})|_D.$$

The main results of these sections are tools for dealing with *multiple modifications*

$$\mathcal{E}[D_1 \to \mathcal{F}_1][D_2 \to \mathcal{F}_2] \cdots [D_n \to \mathcal{F}_n],$$

which correspond to the bundles on X that we would obtain by, say, iteratively applying the construction outlined above; our ability to handle multiple modifications will allow us to inductively degenerate C, peeling off lines one (or sometimes two) at a time. Our study of modifications is divided into two sections: We begin in Chapter 2 by studying modifications of vector bundles on arbitrary varieties; and further study the special case of curves in Chapter 3. This is necessary since we will need to apply results on modifications to the total space of a family of curves.

Our next topic in Chapter 4 is interpolation and its interaction with modifications. For example, under certain conditions we show that if a given vector bundle $\mathcal{E}$, a sub-bundle $\mathcal{F}$, and the quotient $\mathcal{E}/\mathcal{F}$, all satisfy interpolation, then so does the modification $\mathcal{E}[D \to \mathcal{F}]$.

In Chapter 5 and Chapter 6, we respectively define, and calculate, important examples of, certain sub-bundles of normal bundles of curves in projective spaces. These bundles will include the bundle $\mathcal{L}$ appearing in Eq. (1.3), as well as the necessary generalizations thereof (which are necessary, say, when L meets X at two points instead of just one).

In Chapter 7, we prove the necessary ingredients to degenerate C to a reducible curve (e.g. we prove that the conclusion of Theorem 1.2 is an open condition in the Hilbert scheme parameterizing curves of degree d and genus g in $\mathbb{P}^r$).

The heart of the paper is Chapter 8, where all of the previous work enables us to carry out the analysis described above (c.f. Eq. (1.3)). We consider not only the case where L meets X once, but also where X meets L twice, as well as several other variants (e.g. X contained in a hyperplane, simultaneously adding two lines, etc.).

We then move forward with our inductive argument: First, in Chapter 9, we define a certain class of modifications of normal bundles of curves which we will inductively study. Then in Chapter 10, we show how the results of Chapter 8 allow us to reduce interpolation for certain cases of this class of modifications of normal bundles to other "simpler" cases. In Chapter 11, we directly prove that certain modified normal bundles satisfy interpolation; these form the base of our inductive argument.

To finish the proof, we need an intricate combinatorial argument to show that the collection of inductive arguments of Chapter 10, together with the base cases of Chapter 11, imply Theorem 1.2. This is briefly summarized in Chapter 12, and detailed in Appendix A

Finally, in Chapter 13, we further explore the three exceptional cases occurring in Theorem 1.2, understanding geometrically why curves of degree $r+2$ and genus 2 in $\mathbb{P}^r$ do not satisfy interpolation for $r \in \{3, 4, 5\}$. The reason is essentially that the sub-bundle $N_{C/S}$ has too many sections, where S is the surface obtained by taking the union of all lines joining pairs of points $\{p, q\} \subset C$ which are conjugate under the hyperelliptic involution. Using this construction, we also establish Theorem 1.3.

Conventions. Unless otherwise noted, we will consistently make the following conventions.

- We will work over an algebraically closed field K of characteristic 0.
- All varieties are reduced, separated, finite type schemes over K.
- All curves are connected and locally complete intersection (lci); all families of curves have connected lci fibers.
- All vector bundles are locally free sheaves of finite constant rank.
- A subbundle refers to a vector subbundle with locally free quotient.
- All divisors are Cartier.
- We will call a vector bundle nonspecial if it has no higher cohomology.

CHAPTER 2

Elementary modifications in arbitrary dimension

Elementary modifications of vector bundles are a classical topic. Most sources focus on reduced divisors over curves, but the applications we have in mind require to us to relax these hypotheses. The goal of this chapter is to define an appropriate notion of modification and develop its properties.

Let X be a variety and E a vector bundle on it. Given an effective Cartier divisor $D \subset X$ and a subbundle $F \subset E|_U$ defined over an open U containing the support of D, we consider the composition

$$E \longrightarrow E|_D \longrightarrow (E/F)|_D$$

of the restriction to D followed by a quotient. Both parts are surjective, hence so is the composition. We will call the kernel of the composition the *(elementary) modification of E at D along F* and denote it by $E[D \to F]$. Our notation is inspired by the fact that sections of $E[D \to F]$ can be identified with sections of E which point along F when restricted to D:

$$\mathrm{H}^0(E[D \to F]) = \{\sigma \in \mathrm{H}^0(E) \mid \sigma|_D \in \mathrm{H}^0(F|_D)\}.$$

The defining exact sequence of a modification $E[D \to F]$ is

$$0 \longrightarrow E[D \to F] \longrightarrow E \longrightarrow (E/F)|_D \longrightarrow 0. \tag{2.1}$$

The inclusion $E[D \to F] \to E$ becomes an isomorphism when restricted to the complement $X \setminus \mathrm{Supp}(D)$. This is true since the cokernel $(E/F)|_D$ is supported on D.

There is a second sequence, which can also be very handy:

$$0 \longrightarrow E(-D) \longrightarrow E[D \to F] \longrightarrow F|_D \longrightarrow 0. \tag{2.2}$$

This is a consequence of the Snake Lemma applied to the following diagram with exact rows.

$$\begin{array}{ccccccccc}
0 & \longrightarrow & 0 & \longrightarrow & E & = & E & \longrightarrow & 0 \\
 & & \downarrow & & \downarrow & & \downarrow & & \\
0 & \longrightarrow & F|_D & \longrightarrow & E|_D & \longrightarrow & (E/F)|_D & \longrightarrow & 0
\end{array}$$

The third useful sequence

$$0 \longrightarrow F \longrightarrow E[D \to F]|_U \longrightarrow (E|_U/F)(-D) \longrightarrow 0 \tag{2.3}$$

is a corollary of the following diagram.

$$\begin{array}{ccccccccc}
0 & \longrightarrow & 0 & \longrightarrow & E|_U & = & E|_U & \longrightarrow & 0 \\
 & & \downarrow & & \downarrow & & \downarrow & & \\
0 & \longrightarrow & (E|_U/F)(-D) & \longrightarrow & E|_U/F & \longrightarrow & (E|_U/F)|_D & \longrightarrow & 0
\end{array}$$

Note that sequences (2.1) and (2.2) are valid over the entire variety X, while (2.3) makes sense only over the open U.

REMARK 2.1. The inclusion $F \to E[D \to F]|_U$ in (2.3) splits if the inclusion $F \to E|_U$ splits. We can then write

$$E[D \to F]|_U = F \oplus (E|_U/F)(-D).$$

In particular, if U is affine, then both inclusions split.

REMARK 2.2. The modification $E[D \to F]$ only depends on the restriction of F to D. Put differently, if F and F' are subbundles of E such that $F|_D = F'|_D$, then

$$E[D \to F] = E[D \to F'].$$

For example, if the support of D is an irreducible variety $Y \subset X$ and $D = nY$, then $E[D \to F]$ only depends on F in an n-th order neighborhood of Y.

Under the hypotheses we made, elementary modifications are vector bundles.

PROPOSITION 2.3. *If $F \subset E|_U$ is a subbundle and D is a Cartier divisor on X, then $E[D \to F]$ is a vector bundle.*

PROOF. Since $E[D \to F]|_{X\setminus\mathrm{Supp}(D)} \cong E|_{X\setminus\mathrm{Supp}(D)}$, we can pass to an open neighborhood of D. For example, take the locus U where F is defined.

Note that $E[D \to F]$ is finitely presented and it suffices to show it is flat. We will use the local criterion of flatness. Let $\mathcal{A}$ be a coherent sheaf over X. If we apply $\mathcal{T}\mathrm{or}_\bullet(-, \mathcal{A})$ to sequence (2.3), then $\mathcal{T}\mathrm{or}_1(E[D \to F], \mathcal{A})$ sits between $\mathcal{T}\mathrm{or}_1(F, \mathcal{A}) = 0$ and $\mathcal{T}\mathrm{or}_1((E/F)(-D), \mathcal{A}) = 0$, so it must also be zero. This proves that $E[D \to F]$ is flat, hence locally free. □

EXAMPLE 2.4. Taking $F = 0$ and $F = E$ gives rise to two basic examples:

$$E[D \to 0] = E(-D), \qquad E[D \to E] = E.$$

Similarly, when $D = \emptyset$, then

$$E[D \to F] = E$$

is independent of F.

Consider a morphism of varieties $f\colon Y \to X$. Let D be an effective Cartier divisor on X such that its support contains no component of the image of f. Under this hypothesis, the pullback divisor f^*D is well-defined and modifications respect pullbacks.

PROPOSITION 2.5. *Let $f\colon Y \to X$ be a morphism of varieties and D an effective divisor on X such that its support does not contain any component of the image of f. If E is a vector bundle on X and $F \subset E$ a subbundle, then there is a natural isomorphism*

$$f^*E[D \to F] \cong (f^*E)[f^*D \to f^*F].$$

PROOF. This follows by pulling back the defining sequence (2.1). □

REMARK 2.6. Any vector bundle can be decomposed as a gluing of bundles over affine opens which themselves intersect in affines. The compatibility of modifications and pullbacks, open embeddings in particular, allows us to reduce various statements about modifications over general varieties to statements about affine varieties. We will use this technique in several of the arguments that follow.

REMARK 2.7. Suppose we have a vector bundle E over a variety X and a collection of subbundles $F_i \in E$ indexed by $i \in I$. Stating that $\{F_i\}$ are linearly independent means that for all $x \in X$ the fibers $\{F_i|_x\}$ are linearly independent in $E|_x$ as vector spaces.

There is an alternative formulation of this statement. The individual inclusions $F_i \to E$ induce a morphism

$$\varphi \colon \bigoplus_{i \in I} F_i \longrightarrow E.$$

Then $\{F_i\}$ are linearly independent if and only if φ is injective and has locally free cokernel, that is, $\bigoplus_{i \in I} F_i \subset E$ is a subbundle. This restatement is convenient since it allows us to deal with linear independence in a global fashion.

There are correspondences between certain classes of subbundles of E and $E[D \to F]$. If X is a curve this is true more generally without any restrictions on the subbundles in consideration (see Chapter 3).

To state our result, consider a subbundle $F \subset E|_U$ and an effective divisor D on X whose support is contained in U. We define four sets of subbundles of E:

$$\begin{aligned} S_1(E,F,D) &= \{G \subset E \text{ subbundle} \mid G|_V \subset F|_V \\ &\qquad\qquad \text{for some neighborhood } V \subset U \text{ of } D\}, \\ S_2(E,F,D) &= \{G \subset E \text{ subbundle} \mid F|_V \subset G|_V \\ &\qquad\qquad \text{for some neighborhood } V \subset U \text{ of } D\}, \\ S_3(E,F,D) &= \{G \subset E \text{ subbundle} \mid G|_D \text{ and } F|_D \text{ are linearly independent}\}, \text{ and} \\ S(E,F,D) &= S_1(E,F) \cup S_2(E,F,D) \cup S_3(E,F,D). \end{aligned}$$

Direct inspection shows that

$$S_1(E,F,D) \cap S_2(E,F,D) = \{G \subset E \text{ subbundle} \mid G|_V = F|_V \text{ for some neighborhood } V \subset U \text{ of } D\},$$

$$S_1(E,F,D) \cap S_3(E,F,D) = \begin{cases} \{G \subset E \text{ subbundle}\} & \text{if } D = \emptyset, \\ \{0\} & \text{otherwise, and} \end{cases}$$

$$S_2(E,F,D) \cap S_3(E,F,D) = \begin{cases} \{G \subset E \text{ subbundle}\} & \text{if } F|_D = 0, \\ \emptyset & \text{otherwise.} \end{cases}$$

In particular, if $D \cap X' \neq \emptyset$ for every irreducible component $X' \subset X$, then

$$S_1(E,F,D) \cap S_2(E,F,D) = \begin{cases} \{\overline{F}\} & \text{if } F \text{ extends to } \overline{F} \text{ defined over } X, \\ \emptyset & \text{otherwise, and} \end{cases}$$

$$S_1(E,F,D) \cap S_3(E,F,D) = \{0\}.$$

PROPOSITION 2.8. *Let $F \subset E|_U$ be a subbundle and D an effective divisor on X whose support is contained in U. Note that we can also treat F as a subbundle of the modification $E[D \to F]|_U$ by sequence (2.3). Then there are bijections*

$$\begin{aligned}
\varphi_1 \colon S_1(E, F, D) &\longrightarrow S_1(E[D \to F], F, D),\\
\varphi_2 \colon S_2(E, F, D) &\longrightarrow S_2(E[D \to F], F, D),\\
\varphi_3 \colon S_3(E, F, D) &\longrightarrow S_3(E[D \to F], F, D), \text{ and}\\
\varphi \colon S(E, F, D) &\longrightarrow S(E[D \to F], F, D),
\end{aligned}$$

such that

(a) $\varphi|_{S_i(E,F,D)} = \varphi_i$ *for* $i = 1, 2, 3$,
(b) φ *is compatible with pullbacks,*
(c) *given* $G_1, G_2 \in S(E, F, D)$, *then* $G_1 \subset G_2$ *in a neighborhood of* D *if and only if* $\varphi(G_1) \subset \varphi(G_2)$ *in a neighborhood of* D,
(d) *given* $\{G_i \mid i \in I\} \subset S_1(E, F, D)$, *then* $\{G_i\}$ *are linearly independent along* D *if and only if* $\{\varphi(G_i)\}$ *are linearly independent along* D,
(e) *given* $\{G_i \mid i \in I\} \subset S_3(E, F, D)$, *then* $\{F\} \cup \{G_i\}$ *are linearly independent along* D *if and only if* $\{F\} \cup \{\varphi(G_i)\}$ *are linearly independent along* D, *and*
(f) *if* $D = \emptyset$, *then* $E[D \to F] \cong E$ *induces an identification of* $S(E, F, D)$ *and* $S(E[D \to F], F, D)$ *such that* φ *becomes the identity map.*

PROOF. Without loss of generality, assume that X is connected. If not, we can use the morphisms φ_i and φ defined for each connected component to assemble their global versions.

Let us start by considering φ_1. We will bootstrap our way up by first constructing a simpler bijection $\varphi_1' \colon S_1(E|_U, D) \to S_1(E[D \to F]|_U, D)$. Note that F sits in both $E|_U$ and $E[D \to F]|_U$, so we can send $G \subset F \subset E|_U$ to $G' = G \subset F \subset E[D \to F]|_U$. To verify that this map sends subbundles to subbundles, we observe that given $F \subset E|_U$ is a subbundle, then $G \subset E|_U$ is a subbundle if and only if $G \subset F$ is a subbundle. An analogous statement is true for the inclusions $G' \subset F \subset E[D \to F]|_U$. To define φ_1 in terms of φ_1', it suffices to note that φ_1' is compatible with open embeddings (more generally, it is compatible with pullbacks) and $E[D \to F]|_{X \setminus \mathrm{Supp}(D)} \cong E|_{X \setminus \mathrm{Supp}(D)}$. Simply put, the image $G' = \varphi_1(G)$ is glued from $\varphi_1'(G|_U)$ and $G|_{X \setminus \mathrm{Supp}(D)}$ along $U \setminus \mathrm{Supp}(D)$. It is easy to see that both φ_1' and φ_1 are bijections.

The compatibility with pullbacks follows from the fact sequences (2.1) to (2.3) are preserved by pullbacks as long as the hypotheses of Theorem 2.5 are satisfied.

The second and third morphisms, while still induced by the inclusion $E[D \to F] \to E$, are a little more interesting. For example, the main issue with φ_2 is that given a subbundle $G' \subset E[D \to F]$, its image in E is no longer a subbundle. One way to solve the problem is via saturation, but this makes it hard to understand the resulting subbundle $G \subset E$.

Consider the second morphism φ_2. Given a subbundle $G \subset E$ which contains F, we can show that the inclusion $G \to E$ lifts to an inclusion $G[D \to F] \to E[D \to F]$.

We have constructed the following diagram with exact rows.

$$\begin{array}{ccccccccc} 0 & \longrightarrow & G[D \to F] & \longrightarrow & G & \longrightarrow & (G/F)|_D & \longrightarrow & 0 \\ & & \downarrow{\scriptstyle \iota} & & \downarrow & & \downarrow & & \\ 0 & \longrightarrow & E[D \to F] & \longrightarrow & E & \longrightarrow & (E/F)|_D & \longrightarrow & 0 \end{array}$$

All vertical maps are injective, and the Snake Lemma produces the short exact sequence

$$0 \longrightarrow \operatorname{Coker} \iota \longrightarrow E/G \longrightarrow (E/G)|_D \longrightarrow 0,$$

which identifies

$$\operatorname{Coker} \iota \simeq (E/G)(-D).$$

It follows that the inclusion $G[D \to F] \to E[D \to F]$ is a subbundle since we identified its cokernel with $(E/G)(-D)$. The morphism φ_2 can thus be defined by sending G to $G[D \to F]$.

To construct the backward direction of φ_2, we start with $G' \subset E[D \to F]$ which contains F. Similarly to (2.2), there is a morphism $E \to E[D \to F](D)$. We define $G = \varphi_2^{-1}(G')$ as the kernel of the composition

$$E \longrightarrow E[D \to F](D) \longrightarrow E[D \to F](D)/G'(D),$$

which is a subbundle of E. Following our construction, it is easy to check that φ_2 is a bijection and it is compatible with pullbacks.

We proceed to construct φ_3. Let us start by constructing the forward direction first. Take the composition

$$G \longrightarrow E \longrightarrow (E/F)|_D,$$

where $G \subset E$ is such that $G|_D$ and $F|_D$ are linearly independent. By first restricting to D, we can identify its image $((F+G)/F)|_D$ with $G|_D$, so we obtain a morphism

$$G(-D) = \operatorname{Ker}(G \to G|_D) \longrightarrow E[D \to F].$$

Sending G to $G' = G(-D)$ furnishes the forward direction of the bijection. To see that $G(-D) \subset E[D \to F]$ is a subbundle, observe that we have constructed the following diagram with exact rows.

$$\begin{array}{ccccccccc} 0 & \longrightarrow & G(-D) & \longrightarrow & G & \longrightarrow & (F|_D + G|_D)/F|_D & \longrightarrow & 0 \\ & & \downarrow{\scriptstyle \iota} & & \downarrow & & \downarrow & & \\ 0 & \longrightarrow & E[D \to F] & \longrightarrow & E & \longrightarrow & (E/F)|_D & \longrightarrow & 0 \end{array}$$

Applying the Snake Lemma, we obtain the short exact sequence

$$0 \longrightarrow \operatorname{Coker} \iota \longrightarrow E/G \longrightarrow (E/(F+G))|_D \longrightarrow 0,$$

which identifies

$$\operatorname{Coker} \iota \simeq (E/G)[D \to (F+G)/G].$$

In particular, the cokernel of the first vertical map is a vector bundle by Theorem 2.3. A similar analysis constructs the backward direction of the second map which sends $G' \subset E[D \to F]$ to $G = G'(D) \subset E$. Again, all diagrams are preserved by appropriate pullbacks (see Theorem 2.5).

To construct φ it suffices to note that φ_i agree on all pairwise intersections of their domains. This also ensures part (a) is true. On a similar note, part (f) follows immediately from the constructions of φ_i.

Next, we focus on part (c). Without loss of generality, we may replace X with an irreducible component which intersects D non-trivially. Since E and $E[D \to F]$ are isomorphic over $X \setminus \mathrm{Supp}(D)$, in particular, they are isomorphic over the generic point $\eta \in X$. What is more interesting is that if we identify $E|_\eta$ and $E[D \to F]|_\eta$, then φ_1, φ_2, φ_3, and φ become the identity map. Since containment is a closed property, part (c) follows immediately from the observations we made.

We are left to demonstrate parts (d) and (e) of our claim. Consider a subset $\{G_i \mid i \in I\} \subset S_1(E, F, D)$. Assume that $\{G_i\}$ are linearly independent in a neighborhood $V \subset U$ of D. We can replace X with V so $\{G_i\}$ are linearly independent everywhere. Note that F sits in both E and $E[D \to F]$ as a subbundle. Furthermore, all subbundles of F remain unchanged by φ. Since all G_i are contained in F, part (d) follows immediately.

Finally, consider a subset $\{G_i \mid i \in I\} \subset S_3(E, F, D)$ such that $\{F\} \cup \{G_i\}$ are linearly independent in a neighborhood V of D. Recall that $\varphi_3(G_i) = G_i(-D) \subset E[D \to F]$. After replacing X with V, we have a subbundle $F \oplus G \to E$ where $G = \bigoplus_i G_i$. This inclusion lifts to a morphism $G(-D) \to E[D \to F]$ which fits in the following diagram with exact rows.

$$\begin{array}{ccccccccc}
0 & \longrightarrow & F \oplus G(-D) & \longrightarrow & F \oplus G & \longrightarrow & G|_D & \longrightarrow & 0 \\
 & & \downarrow & & \downarrow & & \downarrow & & \\
0 & \longrightarrow & E[F \to D] & \longrightarrow & E & \longrightarrow & (E/F)|_D & \longrightarrow & 0
\end{array}$$

All vertical morphisms are injective, so the Snake Lemma identifies the cokernel of $F \oplus G(-D) \to E[D \to F]$ with

$$(E/(F+G))(-D)$$

which is a vector bundle. We have thus shown that $\{F\} \cup \{\varphi(G_i)\}$ are linearly independent in $E[D \to F]$. The backward implication has an analogous proof, so we will omit that. □

Our discussion so far has only handled single modifications. This is insufficient for our purposes, and we would like to be handle more than one modification at a time. If the underlying variety X is a curve, there is a recursive definition which utilizes the curve-to-projective extension theorem [**7**, I.6.8] and works in full generality (see Chapter 3). In higher dimensions, one needs to be much more careful. The following notions formalize multi-modifications. Later, we will relate these to the recursive definition for curves.

DEFINITION 2.9. Let $\{F_i \subset E \mid i \in I\}$ be a collection of subbundles. We will say that $\{F_i\}$ is *tree-like* at a point $x \in X$ if for all $I' \subset I$ either

(a) the set of subspaces $\{F_i|_x \mid i \in I'\}$ is linearly independent in E_x, or
(b) there is a distinct pair $i, j \in I'$ and an open $U \subset X$ containing $x \in X$ such that $F_i|_U \subset F_j|_U$.

We will use $\mathrm{TL}_X(\{F_i\})$ to denote the set of tree-like points in X. When there is no ambiguity, we may write $\mathrm{TL}(\{F_i\}) = \mathrm{TL}_X(\{F_i\})$. We will say that $\{F_i\}$ is tree-like along $Y \subset X$ if $Y \subset \mathrm{TL}(\{F_i\})$.

REMARK 2.10. Note that being tree-like is a local property. Let $U \subset X$ be an open, and $x \in U$ is a point in it. Being local means that $\{F_i\}$ is tree-like at x if and only if $\{F_i|_U\}$ is tree-like at x. More strongly, being tree-like is also preserved by pullbacks.

On a similar note, since linear independence is an open property, then being tree-like is also open.

REMARK 2.11. The definition of being tree-like is inspired by the following observation. Let E be a vector bundle over a variety X, and $\{F_i \subset E\}$ a collection of subbundles. In addition, we consider the inclusion graph of $\{F_i\} \cup \{E\}$. The collection $\{F_i\}$ is tree-like over X if and only if the following two conditions are satisfied:

(a) the inclusion graph is a tree, and
(b) the children of each node are linearly independent.

The definition of tree-like was crafted so we can transfer multiple subbundles and entire modification data through modifications, similarly to Theorem 2.8. To simplify the statement of the following result, set

$$S^{\text{set}}(E, F, D) = \{\{F_i \subset E \text{ subbundle}\} \mid \{F\} \cup \{F_i\} \text{ is tree-like along } D\}.$$

PROPOSITION 2.12. *Let $F \subset E|_U$ be a subbundle and D an effective divisor on X whose support is contained in U. Then there is a bijection*

$$\varphi^{\text{set}} \colon S^{\text{set}}(E, F, D) \longrightarrow S^{\text{set}}(E[D \to F], F, D)$$
$$\{F_i\} \longmapsto \{\varphi(F_i)\}$$

such that

(a) *φ^{set} is compatible with pullbacks, and*
(b) *if $D = \emptyset$ and we identify $S^{\text{set}}(E, F, D)$ and $S^{\text{set}}(E[D \to F], F, D)$, then φ^{set} becomes the identity map.*

PROOF. As long as we show that φ^{set} is a well-defined bijection, then parts (a) and (b) follow from Theorem 2.8.

Consider a collection of subbundles $\{F_i \subset E \mid i \in I\}$ such that $\{F\} \cup \{F_i\}$ is tree-like along D. For convenience, set $F_0 = F$ and $\overline{I} = \{0\} \sqcup I$. To verify that φ^{set} is well-defined, we first need to show that all F_i are in $S(E, F, D)$, the domain of φ. Fix an index i, and take $I' = \{0, i\}$. By the definition of being tree-like, we know that one of the following is true:

(1) $F_i \subset F$ in a neighborhood of D,
(2) $F \subset F_i$ in a neighborhood of D, or
(3) F and F_i are linearly independent along D.

These cases correspond to $S_1(E, F, D)$, $S_2(E, F, D)$, and $S_3(E, F, D)$ respectively, so $F_i \in S(E, F, D)$. Next, we need to show that $\{F\} \cup \{\varphi(F_i)\}$ is tree-like along D. This follows immediately from the fact that φ respects inclusions and linear independence.

We have demonstrated that φ^{set} is a well-defined map. To conclude our proof, we need to demonstrate it is a bijection. It suffices to note we can construct an inverse $(\varphi^{\text{set}})^{-1}(\{F_i'\}) = \{\varphi^{-1}(F_i')\}$. □

After establishing transfer for sets of subbundles, the next step in our bootstrapping program is to define modification data and show how to transfer them.

DEFINITION 2.13. A *modification datum for E* is an ordered collection of triples

$$M = \{(D_i, U_i, F_i) \mid i \in I\}$$

such that for each i:

(a) D_i is an effective Cartier divisor on X,
(b) $U_i \subset X$ is an open containing the support of D_i, and
(c) $F_i \subset E|_{U_i}$ is a subbundle

In addition, we will call a datum M *tree-like* if for all subsets $I' \subset I$, there is an inclusion

$$\bigcap_{i \in I'} \operatorname{Supp}(D_i) \subset \mathrm{TL}_{U_{I'}}(\{F_i|_{U_{I'}} \mid i \in I'\}),$$

where $U_{I'} = \bigcap_{i \in I'} U_i$. Put differently, for all $x \in X$ the collection of subbundles $\{F_i \mid x \in D_i\}$ is tree-like at x.

To simplify the transfer statement for modification data, set

$$S^{\mathrm{md}}(E, F, D) = \{M = \{(D_i, U_i, F_i)\} \mid \{(D, U, F)\} \cup M \text{ is a tree-like modification datum}\}.$$

PROPOSITION 2.14. *Let $F \subset E|_U$ be a subbundle and D an effective divisor on X whose support is contained in U. Then there is a bijection*

$$\begin{aligned} \varphi^{\mathrm{md}} \colon S^{\mathrm{md}}(E, F, D) &\longrightarrow S^{\mathrm{md}}(E[D \to F], F, D) \\ \{(D_i, U_i, F_i)\} &\longmapsto \{(D_i, U_i, \varphi(F_i))\} \end{aligned}$$

such that

(a) *φ^{md} is compatible with pullbacks, and*
(b) *if $D = \emptyset$ and we identify $S^{\mathrm{md}}(E, F, D)$ and $S^{\mathrm{md}}(E[D \to F], F, D)$, then φ^{md} becomes the identity map.*

PROOF. Continuing our build-up, we will repeatedly refer to Theorem 2.12 in this proof. First, parts (a) and (b) follow immediately once we establish that φ^{md} is a well-defined bijection.

Fix an element $M = \{(D_i, U_i, F_i) \mid i \in I\} \in S^{\mathrm{md}}(E, F, D)$. As before, we set

$$F_0 = F, \qquad U_0 = U, \qquad D_0 = D, \qquad \overline{I} = \{0\} \cup I, \qquad \overline{M} = \{(D, U, F)\} \cup M.$$

Given a subset $I' \subset \overline{I}$, we know that the intersection $\bigcap_{i \in I'} D_i$ lies in the set of tree-like points

$$V_{I'} = \mathrm{TL}_{U_{I'}}(\{F_i \mid i \in I'\}).$$

Applying Theorem 2.12 to $\{F_i|_{V_I'} \mid i \in I'\}$, we conclude that

$$V_{I'} = \mathrm{TL}_{U_{I'}}(\{\varphi(F_i) \mid i \in I'\}).$$

We have demonstrated that $\{(D, U, F)\} \cup \{(D_i, U_i, \varphi(F_i))\}$ is a tree-like modification datum, so $\varphi^{\mathrm{md}}(M) \in S^{\mathrm{md}}(E[D \to F], F, D)$ and φ^{md} is a well-defined map. To see that it is a bijection, it suffices to note we can construct an inverse using $(\varphi^{\mathrm{set}})^{-1}$. □

We are now ready to provide a general definition of vector bundle modifications. The main idea is to recursively use the transfer of modification data (Theorem 2.14).

DEFINITION 2.15. Let X be a variety, E a vector bundle over X, and M a tree-like modification datum for E. If M is empty, then we define $E[\emptyset] = E$. On the other hand, if $M = \{(D, U, F)\} \cup M'$, then

$$E[M] = E[D \to F][\varphi^{\mathrm{md}}(M')],$$

where $\varphi^{\mathrm{md}} \colon S^{\mathrm{md}}(E, F, D) \to S^{\mathrm{md}}(E[D \to F], F, D)$ is the transfer map described in Theorem 2.14. When

$$M = \{(D_1, U_1, F_1), \dots, (D_m, U_m, F_m)\},$$

we will allow ourselves to write

$$E[M] = E[D_1 \to F_1] \cdots [D_m \to F_m].$$

After establishing the language of multi-modifications, we are ready to describe some of its basic properties. First, we note that modifications respect pullbacks. This is a direct consequence of Theorem 2.5.

COROLLARY 2.16. *Let $f \colon Y \to X$ be a morphism of varieties and E a vector bundle on X. If $M = \{(D_i, U_i, F_i)\}$ is a tree-like modification datum for E such that $\bigcup_i \operatorname{Supp}(D_i)$ does not contain any component of the image of f, then the pullback datum*

$$f^*M = \{(f^*D_i, f^{-1}(U_i), f^*F_i)\}$$

is tree-like, and there is a natural isomorphism

$$f^*E[M] \cong (f^*E)[f^*M].$$

Next, note that we defined a modification datum as an ordered collection of triples (see Theorem 2.13). While the order plays a crucial point in our formulation, it turns out to be irrelevant for the final result $E[M]$ as long as M is a tree-like modification datum.

PROPOSITION 2.17 (Commuting modifications). *Let E be a vector bundle over a variety X, and M a tree-like modification datum. If M' is a datum obtained by reordering M, then there is a natural isomorphism $E[M] \cong E[M']$ compatible with pullbacks.*

PROOF. Since any symmetric group is generated by transpositions, it suffices to consider the case

$$M = \{(D_1, U_1, F_1), (D_2, U_2, F_2)\}, \qquad M' = \{(D_2, U_2, F_2), (D_1, U_1, F_1)\}.$$

We also need to know that $\varphi^{\mathrm{md}}_{F_1,D_1} \circ \varphi^{\mathrm{md}}_{F_2,D_2} = \varphi^{\mathrm{md}}_{F_2,D_2} \circ \varphi^{\mathrm{md}}_{F_1,D_1}$ for the subset of the domain where this composition makes sense. If we assume there is an isomorphism $E[M] \cong E[M']$, this statement is automatically true if we pass to any the generic point. But subbundles which agree on all generic points must be the same, so this issue is resolved.

We proceed by making several reductions. First, there is a natural isomorphism $E[M] \cong E[M']$ over $X \setminus (\operatorname{Supp}(D_1) \cap \operatorname{Supp}(D_2))$, so it suffices to focus on a neighborhood of $\operatorname{Supp}(D_1) \cap \operatorname{Supp}(D_2)$. Next, we can cover this locus by affine opens U which fall in one of the following three categories: (1) $F_1|_U \subset F_2|_U$, (2) $F_2|_U \subset F_1|_U$, or (3) F_1 and F_2 are linearly independent over U. Since cases (1) and (2) are analogous, so we will demonstrate (1) and (3). For simplicity, we can also replace X with U.

Assume that $F_1 \subset F_2$. Since we are working over an affine space, there are splittings $F_2 = F_1 \oplus F_1'$ and $E = F_2 \oplus F_2'$. Then

$$\begin{aligned} E &= F_1 \oplus F_1' \oplus F_2', \\ E[D_1 \to F_1] &= F_1 \oplus F_1'(-D_1) \oplus F_2'(-D_1), \text{and} \\ E[D_2 \to F_2] &= F_1 \oplus F_1' \oplus F_2'(-D_2). \end{aligned}$$

Using these splittings, we can perform the second modification to arrive at

$$\begin{aligned} E[M] &= E[D_1 \to F_1][D_2 \to F_2] \\ &= F_1 \oplus F_1'(-D_1) \oplus F_2'(-D_1 - D_2) \\ &= E[D_2 \to F_2][D_1 \to F_1] \\ &= E[M']. \end{aligned}$$

In the third case, we assume F_1 and F_2 are linearly independent which leads to a splitting $E = F_1 \oplus F_2 \oplus F$. A similar computation demonstrates that

$$\begin{aligned} E[D_1 \to F_1] &= F_1 \oplus F_2(-D_1) \oplus F(-D_1), \text{and} \\ E[D_2 \to F_2] &= F_1(-D_2) \oplus F_2 \oplus F(-D_2), \end{aligned}$$

and

$$E[M] = F_1(-D_2) \oplus F_2(-D_1) \oplus F(-D_1 - D_2) = E[M']. \qquad \square$$

PROPOSITION 2.18 (Commuting modifications and twists). *Let E be a vector bundle over a variety X, $F \subset E$ a subbundle, and $M = \{(D_i, U_i, F_i)\}$ a tree-like modification datum. If D is a Cartier divisor (not necessarily effective) and we define the datum $M(D) = \{(D_i, U_i, F_i(D))\}$ for $E(D)$, then $M(D)$ is tree-like and there is a natural isomorphism*

$$E[M](D) = E(D)[M(D)]$$

compatible with pullbacks.

PROOF. To see that $M(D)$ is tree-like, it suffices to note that vector bundle inclusion and linear independence are preserved by twisting.

First, assume we know the desired isomorphism exists for negative effective divisors. Given a divisor D, we can always decompose it as $D = D^+ - D^-$ where D^+ and D^- are effective. Using the pair $E(D^+)$ and $M(D^+)$ with divisor $-D^+$, we deduce

$$\begin{aligned} E[M] &= E(D^+ - D^+)[M(D^+ - D^+)] \\ &\cong E(D^+)[M(D^+)](-D^+). \end{aligned}$$

Next, we apply the same result for $E(D^+)$, $M(D^+)$ with divisor $-D^-$:

$$\begin{aligned} E[M](D) &= E[M](D^+ - D^-) \\ &\cong E(D^+)[M(D^+)](-D^-) \\ &\cong E(D^+ - D^-)[M(D^+ - D^-)] \\ &= E(D)[M(D)]. \end{aligned}$$

We are left to furnish an isomorphism in the case of negative effective divisors. For simplicity, replace D with its negative, so it is effective. Let U be a neighborhood of $\operatorname{Supp} D$. Note that if M is a tree-like datum, then $M' = M \cup \{(D, U, 0)\}$ is also tree-like. Since $E[D \to 0] \cong E(-D)$, then the associated morphism φ^{md} maps the datum M to $M(-D)$. Commutativity implies

$$\begin{aligned} E[M](-D) &\cong E[M][D \to 0] \\ &\cong E[M'] \\ &\cong E[D \to 0][\varphi^{\mathrm{md}}(M)] \\ &\cong E(-D)[M(-D)], \end{aligned}$$

which concludes our argument. □

REMARK 2.19. When it is clear that M is a modification datum for E, we will allow ourselves to write M instead of $M(D)$. Then the statement of Theorem 2.18 becomes

$$E[M](D) = E(D)[M],$$

so we say that modifications and twists commute.

If we focus on the case of two modifications with identical base divisors, there are two more results mentioning.

PROPOSITION 2.20 (Combining modifications). *Let E be a vector bundle over a variety X. Consider a tree-like modification datum $M = \{(aD, U, F_1), (bD, U, F_2)\}$ for E, where a, b is a pair of non-negative integers.*

(a) *If $F = F_1 = F_2$, then*

$$E[aD \to F][bD \to F] \cong E[(a+b)D \to F].$$

(b) *If F_1, F_2 are linearly independent and $a = b = 1$, then*

$$E[D \to F_1][D \to F_2] \cong E[D \to F_1 + F_2](-D).$$

In addition, both isomorphisms are compatible with pullbacks.

PROOF. Following Theorem 2.6, we can assume X is affine. For part (a), there is a splitting $E = F \oplus E/F$, and we compute

$$\begin{aligned} E[aD \to F][bD \to F] &\cong (F \oplus (E/F)(-aD))[bD \to F] \\ &\cong F \oplus (E/F)(-aD)(-bD) \\ &\cong F \oplus (E/F)(-(a+b)D) \\ &\cong E[(a+b)D \to F]. \end{aligned}$$

In part (b), consider a splitting $E = F_1 \oplus F_2 \oplus F_3$. Then

$$\begin{aligned} E[D \to F_1][D \to F_2] &\cong (F_1 \oplus F_2(-D) \oplus F_3(-D))[D \to F_2(-D)] \\ &\cong F_1(-D) \oplus F_2(-D) \oplus F_3(-2D) \\ &\cong (F_1 \oplus F_2 \oplus F_3(-D))(-D) \\ &\cong E[D \to F_1 + F_2](-D). \end{aligned}$$

□

CHAPTER 3

Elementary modifications for curves

While Chapter 2 introduces vector bundle modifications in a very general setting, the applications we have in mind use curves and families of curves. The present chapter will explain more concretely how modifications manifest themselves for curves, and provide several simple consequences.

A substantial part of bootstrapping the definition of multiple modifications consisted of transfer statements. It turns out that curves allow for a simpler transfer statement for subbundles which extends Theorem 2.8. In particular, this allows us to extend multi-modifications beyond tree-like data at the expense of sacrificing some of the properties we already established (e.g., commutativity).

To state our result, define

$$\overline{S}(E) = \{G \subset E \text{ subbundles}\},$$

where E is a vector bundle over a curve C.

PROPOSITION 3.1. *Let E be a vector bundle over a curve C. Given a subbundle $F \subset E$ and a divisor D whose support is contained in the smooth locus of C, there is a bijection*

$$\overline{\varphi}\colon \overline{S}(E) \longrightarrow \overline{S}(E[D \to F]),$$

such that

(a) $\overline{\varphi}|_{S(E,F,D)} = \varphi$ *where $S(E,F,D)$ and φ are as in Theorem 2.8,*
(b) $\overline{\varphi}$ *is compatible with pullbacks,*
(c) *given $G_1, G_2 \in \overline{S}(E)$, then $G_1 \subset G_2$ in a neighborhood of D implies $\overline{\varphi}(G_1) \subset \overline{\varphi}(G_2)$ in a neighborhood of D, and*
(d) *if $D = \emptyset$, then $E[D \to F] \cong E$ induced an identification of $\overline{S}(E)$ and $\overline{S}(E[D \to F])$ such that $\overline{\varphi}$ becomes the identity map.*

PROOF. We start by constructing the map $\overline{\varphi}$. Given a subbundle $G \subset E$ of rank r, we can produce a section σ of the Grassmannian bundle $\mathrm{Gr}(r, E)$ of E.

$$\begin{array}{c} \mathrm{Gr}(r,E) \\ \downarrow \uparrow \sigma \\ C \end{array}$$

The natural inclusion $E[D \to F]$ is an isomorphism over $U = C \setminus \mathrm{Supp}(D)$, so we also have an isomorphism $\mathrm{Gr}(r, E)|_U \cong \mathrm{Gr}(r, E[D \to F])|_U$. It follows that we can treat $\sigma|_U$ as a section of the second Grassmannian bundle over U. The curve-to-projective extension theorem [**7**, I.6.8] implies there is a unique section $\sigma'\colon C \to \mathrm{Gr}(r, E[D \to F])$ which extends $\sigma|_U$. The new section gives rise to a subbundle $\overline{\varphi}(G) = G' \subset E[D \to F]$.

For part (a), start by picking a bundle $G \in S(E, F, D)$. If we identify $E|_U$ and $E[D \to F]|_U$, then $\overline{\varphi}(G)|_U = G|_U = \varphi(G)|_U$. Since both $\varphi(G)$ and $\overline{\varphi}(G)$ are subbundles, and $U \subset C$ is dense, it follows that $\varphi(G) = \overline{\varphi}(G)$.

Note that it makes sense to consider the pullback by a morphism $f\colon C' \to C$ only if the pullback divisor f^*D is well-defined. This happens exactly when no component of C' is contracted to a point which lies in the support of the divisor D on C (see Theorem 2.5). In particular, the condition is always satisfied for finite morphisms f. Once we understand this limitation, running through the section extension definition of $\overline{\varphi}$, it is clear that $\overline{\varphi}$ is compatible with pullbacks.

Finally, the proofs of (c) and (d) are identical to the arguments we gave in Theorem 2.8. □

REMARK 3.2. Note that $\overline{\varphi}$ satisfies all properties φ does except it does not preserve linear dependence and independence. To illustrate the point, take $C = \mathbb{A}^1$ with a coordinate x on it, $p = 0$ is the origin, and $E = \mathcal{O}_C \oplus \mathcal{O}_C$. Set

$$F = \langle(1,1)\rangle, \qquad G_1 = \langle(1,0)\rangle, \qquad G_2 = \langle(0,1)\rangle,$$
$$F' = \langle(1,0)\rangle, \qquad G_1' = \langle(1,x)\rangle, \qquad G_2' = \langle(1,-x)\rangle.$$

Then G_1 and G_2 are linearly independent in E, while $\overline{\varphi}(G_1)$ and $\overline{\varphi}(G_2)$ coincide over p in $E[p \to F]$. On the other hand, G_1' and G_2' are linearly dependent at p, but their transfers $\overline{\varphi}(G_1'), \overline{\varphi}(G_2') \subset E[p \to F']$ are linearly independent at p.

In summary, it is possible to modify curves along modification data which are not tree-like, but we need to be careful about switching the order of modifications. Unless otherwise stated, all modifications will be tree-like.

Finally, we present a result which relates the Euler characteristics of a modified bundle and the original one.

PROPOSITION 3.3 (The Euler characteristic of modifications). *Let E be a vector bundle over a curve C.*

(a) *If $D_1, \ldots, D_m$ are effective divisors, and $F_1, \ldots, F_m \subset E$ are subbundles, then*
$$\chi(E[D_1 \to F_1] \cdots [D_m \to F_m]) = \chi(E) - \sum_{i=1}^{m} \deg(D_i) \operatorname{rank}(E/F_i).$$

(b) *If D is a any divisor, then*
$$\chi(E(D)) = \chi(E) + \operatorname{rank}(E) \deg(D).$$

PROOF. Note that the general statement of part (a) follows by applying the $m = 1$ case several times. When $m = 1$, we take Euler characteristics of the sequence (2.1) and note that
$$\chi((E/F_1)|_{D_1}) = \deg(D_1) \operatorname{rank}(E/F_1).$$

Similarly to the proof of Theorem 2.18, we can reduce (b) to the case of a negative effective divisor which is subsumed by part (a). □

REMARK 3.4. The theory of modifications over general varieties (Chapter 2) is certainly more complicated than the statements we presented for curves. Dimensions greater than one become very useful when we deal with families of curves and vector bundles. The fact that constructing modifications preserve pullbacks allows us to treat a modification over the total space of a family of curves as a family of modifications over the individual curves.

We will demonstrate this point through a simple example. Let C be a smooth curve, E a vector bundle over C, and $F \subset E$ a subbundle. We consider the family of curves

$$\mathrm{pr}_2 \colon \mathcal{C} = C \times B \longrightarrow B$$

where $B = C$. Given $b \in B$, we will use $i_b \colon C \to \mathcal{C}$ to denote the inclusion of the fiber over the point b. Choose a point $p_0 \in C$, and construct the divisors

$$D_0 = \{p_0\} \times B, \qquad D_1 = \Delta_C, \qquad D = D_0 + D_1.$$

If

$$E' = (\mathrm{pr}_1^* E)[D \to \mathrm{pr}_1^* F]$$

is the global modification, then restricting to a fiber over b gives

$$i_b^* E' = E[i_b^* D \to F] = E[(b + p_0) \to F].$$

This shows that varying the modification divisor in a family produces modifications which also fit in a family. Furthermore, we know that $E[2p_0 \to F]$ is the "limit modification" as b approaches p_0. This is a very simple example to illustrate the power of modifications over higher dimensional varieties. In general, understanding limits of multiple modifications can be very tricky and being tree-like is the right condition to back our intuitive notion of limits.

CHAPTER 4

Interpolation and short exact sequences

The goal of this chapter is to define interpolation for vector bundles and develop some of its properties, in particular its behavior in short exact sequences. For a more detailed explanation of this property, see [**1**].

DEFINITION 4.1. Let E be a rank n vector bundle over a curve C. We say that a subspace of sections $V \subseteq \mathrm{H}^0(E)$ satisfies interpolation if E is nonspecial, and for every $d \geq 1$, there exists a collection of d points $p_1, \ldots, p_d \in C_{\mathrm{sm}}$ such that

$$\dim\Big(V \cap \mathrm{H}^0\Big(E\Big(-\sum p_i\Big)\Big)\Big) = \max\{0, \dim V - dn\}.$$

We say that E satisfies interpolation if the full space of sections $V = \mathrm{H}^0(E) \subseteq \mathrm{H}^0(E)$ satisfies interpolation.

There are a number of observations which allow us to verify interpolation more easily.

REMARK 4.2. By the upper semi-continuity of h^0, the existence of d points satisfying the equality above implies that a general collection of d points (in one component of C^d_{sm}) satisfies this condition.

REMARK 4.3. In fact, we do not need to check the interpolation condition for every positive integer d. It suffices to verify that the statement holds for $\lfloor \mathrm{h}^0(E)/n \rfloor$ and $\lceil \mathrm{h}^0(E)/n \rceil$. The first value implies the statement holds for all $d \leq \lfloor \mathrm{h}^0(E)/n \rfloor$ and the second for all values $d \geq \lceil \mathrm{h}^0(E)/n \rceil$.

We have arrived at a convenient rephrasing of Theorem 4.1. Let $\mathrm{h}^0(E) = n{\cdot}d{+}r$ where $0 \leq r < n$. Consider the following two statements.

(a) There exist points $p_1, \ldots, p_d \in C_{\mathrm{sm}}$ such that

$$\mathrm{h}^0(E(-\sum p_i)) = r.$$

(b) There exist points $p_1, \ldots, p_{d+1} \in C_{\mathrm{sm}}$ such that

$$\mathrm{h}^0(E(-\sum p_i)) = 0.$$

Assume E has no higher cohomology. If $r = 0$, then interpolation for E is equivalent to (a). In the cases when $r > 0$, interpolation is equivalent to (a) and (b) together.

REMARK 4.4. It is also possible to use the language of divisors to characterize interpolation. Consider a vector bundle $E \to C$ satisfying interpolation. Given an integer $d \geq 1$, there is a component of $\mathrm{Sym}^d C$ so that a general effective divisor D in that component satisfies either $\mathrm{h}^1(E(-D)) = 0$ (when $\deg D \leq \mathrm{h}^0(E)/\operatorname{rank}(E)$) or $\mathrm{h}^0(E(-D)) = 0$ (when $\deg D \geq \mathrm{h}^0(E)/\operatorname{rank}(E)$). Conversely, if for all d there is some component of $\mathrm{Sym}^d C$ for which this disjunction holds, then we can deduce interpolation. We have arrived at the following restatement of Theorem 4.1.

PROPOSITION 4.5. *A nonspecial vector bundle $E \to C$ satisfies interpolation if and only if for every $d \geq 1$, there is a component of $\mathrm{Sym}^d C$ so that a general effective Cartier divisor D of degree d in that component satisfies*

$$\mathrm{h}^0(E(-D)) = 0 \qquad \text{or} \qquad \mathrm{h}^1(E(-D)) = 0.$$

There is a further simplification worth mentioning. Note that we do not need to verify the vector bundle is nonspecial before applying this result.

PROPOSITION 4.6. *A vector bundle E of rank n satisfies interpolation if and only if*

(a) *a general (in some component) effective divisor D of degree $\lceil \mathrm{h}^0(E)/n \rceil$ satisfies $\mathrm{h}^0(E(-D)) = 0$, and*
(b) *a general (in some component) effective divisor D of degree $\lfloor \mathrm{h}^0(E)/n \rfloor$ satisfies $\mathrm{h}^1(E(-D)) = 0$.*

Furthermore, if $\chi(E) \geq 0$, we can replace $\mathrm{h}^0(E)$ with $\chi(E)$ in $\lceil \mathrm{h}^0(E)/n \rceil$ and $\lfloor \mathrm{h}^0(E)/n \rfloor$.

PROOF. To conclude that E is nonspecial, we note that $\mathrm{h}^1(E(-D)) = 0$ for some effective divisor of non-negative degree $\lfloor \mathrm{h}^0(E)/n \rfloor$. The first part is a direct consequence of Theorem 4.5 and Theorem 4.3. For the second part, it suffices to note the same argument implies that $\mathrm{h}^1(E) = 0$ as long as $\chi(E) \geq 0$. □

Characterizing line bundles which satisfy interpolation is particularly simple and worth elaborating on.

PROPOSITION 4.7. *A line bundle satisfies interpolation if and only if it is nonspecial.*

PROOF. One direction is implied by the definition of interpolation. For the converse, consider a nonspecial line bundle L. We proceed to choose $m = \mathrm{h}^0(L)$ points $p_i \in C_{\mathrm{sm}}$ as follows. First, pick p_1 such that $\mathrm{h}^0(L(-p_1)) = \mathrm{h}^0(L) - 1$. If $m \geq 2$, we choose a second point p_2 such that $\mathrm{h}^0(L(-p_1 - p_2)) = \mathrm{h}^0(L) - 2$, and so on. This demonstrates that L satisfies interpolation. □

Next we show interpolation is preserved by modifications along appropriately general subbundles, and by positive twists. To provide the precise statement, we need to introduce the following notion.

DEFINITION 4.8. Let V be a vector space, and $\{W_b \subset V \mid b \in B\}$ be a collection of subspaces indexed by a set B. We will call $\{W_b\}$ *linearly general* if for each subspace $W \subset V$, there exists $b \in B$ such that W_b and W intersect transversely.

REMARK 4.9. Suppose the ambient vector space has dimension n, and all members of the collection $\{W_b\}$ have dimension m. Then to conclude that $\{W_b\}$ is linearly general, it suffices to know that for all subspaces $W \subset V$ of complimentary dimension $n - m$ there exists $b \in B$ such that $W \cap W_b = 0$.

PROPOSITION 4.10. *Let E be a vector bundle over a curve C and $p \in C_{\mathrm{sm}}$ a smooth point. Suppose we have a collection of vector bundles $\{G_b \subset E \mid b \in B\}$ indexed by a set B and $F \subset E$ is a subbundle, such that*

(a) *$F|_p \subset G_b|_p$ for all $b \in B$, and*
(b) *$\{G_b/F|_p \mid b \in B\}$ is linearly general in $E/F|_p$.*

If E and $E[p \to F]$ both satisfy interpolation, then $E[p \to G_b]$ satisfies interpolation for at least one element $b \in B$.

PROOF. We can assemble two copies of sequence (2.1) into the following diagram with exact rows and columns.

$$\begin{array}{ccccccccc}
 & & & & & & 0 & & \\
 & & & & & & \downarrow & & \\
 & & 0 & & & & G_b/F|_p & & \\
 & & \downarrow & & & & \downarrow & & \\
0 & \longrightarrow & E[p \to F] & \longrightarrow & E & \longrightarrow & E/F|_p & \longrightarrow & 0 \\
 & & \downarrow & & \| & & \downarrow & & \\
0 & \longrightarrow & E[p \to G_b] & \longrightarrow & E & \longrightarrow & E/G_b|_p & \longrightarrow & 0 \\
 & & \downarrow & & & & \downarrow & & \\
 & & G_b/F|_p & & & & 0 & & \\
 & & \downarrow & & & & & & \\
 & & 0 & & & & & &
\end{array}$$

Given a divisor D, we twist the entire sequence by $-D$ and take cohomology. For $A = E/F$, E/G_b, and G_b/F, there are induced isomorphisms $A(-D)|_p \cong A|_p$, so we will use the latter. We will also avoid the H^0-functor in front of skyscraper sheaves supported on a point. The operation we described leads to the following diagram.

$$\begin{array}{ccccccccccccc}
 & & & & & & G_b/F|_p & & & & & & \\
 & & & & & & \downarrow & & & & & & \\
0 & \to & \mathrm{H}^0(E[p \to F](-D)) & \to & \mathrm{H}^0(E(-D)) & \to & E/F|_p & \to & \mathrm{H}^1(E[p \to F](-D)) & \to & \mathrm{H}^1(E(-D)) & \to & 0 \\
 & & \downarrow & & \| & & \downarrow & & \twoheaddownarrow & & \| & & \\
0 & \to & \mathrm{H}^0(E[p \to G_b](-D)) & \to & \mathrm{H}^0(E(-D)) & \to & E/G_b|_p & \to & \mathrm{H}^1(E[p \to G_b](-D)) & \to & \mathrm{H}^1(E(-D)) & \to & 0
\end{array}$$

Now that we have described the basic tools we need, we can proceed with the proof. For each $d \geq 1$, choose a divisor D_d of degree d such that

$$\mathrm{h}^0(A(-D_d)) = 0 \qquad \text{or} \qquad \mathrm{h}^1(A(-D_d)) = 0$$

for $A = E$ and $A = E[p \to F]$ (Theorem 4.5). Note that each value d falls in one of three cases:

(1) $\mathrm{h}^0(E(-D_d)) = 0$,
(2) $\mathrm{h}^1(E(-D_d)) = 0$ and $\mathrm{h}^1(E[p \to F](-D_d)) = 0$, or
(3) $\mathrm{h}^1(E(-D_d)) = 0$ and $\mathrm{h}^0(E[p \to F](-D_d)) = 0$.

With the aid of the diagram above, case 1 implies $\mathrm{h}^0(E[p \to G_b](-D_d)) = 0$, and case 2 implies $\mathrm{h}^0(E[p \to G_b](-D_d)) = 0$.

Note that our argument so far works for all $b \in B$. The handling of case 3 requires a choice of b. Fortunately, there can be at most one value of d which satisfies this case. First observe that $\mathrm{H}^0(E(-D_d)) \to E/F|_p$ is an inclusion, and we choose $b \in B$ so that $G_b/F|_p$ is transverse to $\mathrm{H}^0(E(-D))$. It follows that the composition $\mathrm{H}^0(E(-D)) \to E/F|_p \to E/G_b|_p$ has maximal rank. Injectivity and surjectivity respectively imply $\mathrm{h}^0(E[p \to G_b](-D_d)) = 0$ and $\mathrm{h}^1(E[p \to G_b](-D_d)) = 0$. □

PROPOSITION 4.11. *If E satisfies interpolation, and D is any effective Cartier divisor, then $E(D)$ satisfies interpolation.*

PROOF. We need to show that for every degree d, there exists a divisor D' of degree d such that either $\mathrm{h}^0(E(D - D')) = 0$ or $\mathrm{h}^1(E(D - D')) = 0$. If $d > \deg D$, take $D' = D + D''$ such that $\mathrm{h}^0(E(-D'') = 0$ or $\mathrm{h}^1(E(-D'')) = 0$ from the interpolation of E. If $d = \deg D$, take $D' = D$ and note that E is nonspecial. Since $\mathrm{h}^1(E(D - D')) = 0$ is an open condition in D', it follows that there exists some $D' = D_0$ supported on C_{sm} such that $\mathrm{h}^1(E(D - D_0)) = 0$. The interesting case is $d < \deg D$. If we choose an effective divisor $D' \leq D_0$, then $\mathrm{h}^1(E(D - D')) = 0$ follows from $\mathrm{h}^1(E(D - D_0)) = 0$. □

We now study several strengthenings and partial converses to the above results, subject to additional hypotheses — *including the irreducibility of C, which we suppose for the remainder of this section.*

We have already investigated interpolation and twisting up (see Theorem 4.11). The following result provides a partial converse; note that the base curve C needs to be irreducible and $\chi(E)$ is relatively large.

PROPOSITION 4.12. *Let E be a vector bundle on an irreducible curve C, and D an effective divisor on C. If*

(a) *$E(D)$ satisfies interpolation, and*
(b) $\chi(E) \geq \mathrm{genus}(C)\,\mathrm{rank}(E)$,

then E also satisfies interpolation.

PROOF. Since interpolation is an open condition, we may replace D by a divisor supported on the smooth locus of C.

By Theorem 4.6, we only need to show that $\mathrm{h}^1(E(-D')) = 0$ for a general divisor D' of degree $\lfloor \chi(E)/\mathrm{rank}(E) \rfloor$, and $\mathrm{h}^0(E(-D')) = 0$ for general D' of degree $\lceil \chi(E)/\mathrm{rank}(E) \rceil$. Since the arguments are analogous, we will focus on the first case.

For convenience, set $d = \lfloor \chi(E)/\mathrm{rank}(E) \rfloor$ and $g = \mathrm{genus}(C)$. Since $d \geq g$, the Riemann-Roch theorem implies that the natural map $\mathrm{Sym}^d C \to \mathrm{Pic}^d C$ is dominant; hence, it suffices to show that $\mathrm{h}^1(E \otimes L^\vee) = 0$ for a line bundle L of degree d. Since $E(D)$ satisfies interpolation, we know that there exists a divisor D'' of degree $d + \deg(D)$ such that $\mathrm{h}^1(E(D - D'')) = 0$. Taking $L = \mathcal{O}_C(D'' - D)$ completes the argument. □

REMARK 4.13. Suppose we have a family of curves $\pi \colon \mathcal{C} \to B$ whose central fiber $\mathcal{C}_0 = \pi^{-1}(0)$ is reducible but the general fiber is irreducible. If $\mathcal{E}$ is a vector bundle on $\mathcal{C}$ whose restriction $\mathcal{E}_0$ to $\mathcal{C}_0$ satisfies the hypotheses of Theorem 4.12, then the general fiber $\mathcal{C}_b = \pi^{-1}(b)$ together with $\mathcal{E}_b = \mathcal{E}|_{\mathcal{C}_b}$ also satisfy these conditions. While we cannot conclude that $\mathcal{E}_0$ satisfies interpolation, the general bundle $\mathcal{E}_b$ does satisfy interpolation. This will be sufficient for our needs in this paper.

To study the behavior of interpolation under modifications without any assumption of linear generality, we will need to investigate the behavior of interpolation in exact sequences. We begin by considering a short exact sequence

$$0 \longrightarrow F \longrightarrow G \longrightarrow H \longrightarrow 0 \tag{4.1}$$

of vector bundles over an irreducible curve C. Given a Cartier divisor D, we twist back by D and consider associated the long exact sequence in cohomology.

$$\begin{aligned} 0 \longrightarrow \mathrm{H}^0(F(-D)) \longrightarrow \mathrm{H}^0(G(-D)) \longrightarrow \mathrm{H}^0(H(-D)) \xrightarrow{\delta_D} \\ \xrightarrow{\delta_D} \mathrm{H}^1(F(-D)) \longrightarrow \mathrm{H}^1(G(-D)) \longrightarrow \mathrm{H}^1(H(-D)) \longrightarrow 0 \end{aligned} \tag{4.2}$$

We will use $\delta_D\colon \mathrm{H}^0(H(-D)) \to \mathrm{H}^1(F(-D))$ to denote the only non-trivial connecting homomorphism. Our first result allows us to transfer interpolation from the edges F and H to the middle term G.

PROPOSITION 4.14. *Let F, G, and H be as above. If F and H satisfy interpolation, then G satisfies interpolation if and only if*

(a) $\mathrm{h}^0(F)/\operatorname{rank}(F) \le \lfloor \mathrm{h}^0(H)/\operatorname{rank}(H) \rfloor + 1$, *and*
(b) *for every $d \geq 1$ and a general effective divisor D of degree d, the boundary map δ_D has maximal rank (i.e., it is either injective or surjective).*

PROOF. First, assume that G satisfies interpolation. By Theorem 4.5, this means that for a general effective D, either $\mathrm{h}^0(G(-D)) = 0$ or $\mathrm{h}^1(G(-D)) = 0$. Using sequence (4.2), the first case implies that $\mathrm{h}^0(F(-D)) = 0$ and δ_D is injective, and the second that $\mathrm{h}^1(H(-D)) = 0$ and δ_D is surjective. In particular, we have demonstrated condition (b) stating that δ_D has maximal rank.

Condition (a) is a little more interesting. Its negative asserts there exists an integer d such that

$$\frac{\mathrm{h}^0(H)}{\operatorname{rank}(H)} < d < \frac{\mathrm{h}^0(F)}{\operatorname{rank}(F)}. \tag{4.3}$$

For contradiction, assume such an integer d exists. Let D be a general effective divisor of degree d and consider the associated sequence (4.2). The first side of the inequality implies

$$\mathrm{h}^0(H(-D)) = 0 \qquad \text{and} \qquad \mathrm{h}^1(H(-D)) > 0,$$

while the second side implies

$$\mathrm{h}^0(F(-D)) > 0 \qquad \text{and} \qquad \mathrm{h}^1(F(-D)) = 0.$$

Then

$$\mathrm{h}^0(G(-D)) = \mathrm{h}^0(F(-D)) > 0 \qquad \text{and} \qquad \mathrm{h}^1(G(-D)) = \mathrm{h}^1(H(-D)) > 0,$$

which contradicts our hypothesis that G satisfies interpolation by Theorem 4.5.

The reverse direction also follows by inspecting sequence (4.2). Let us pick an integer $d \geq 1$ and consider a general effective divisor D of degree d. We are given that F and H satisfy interpolation, so the argument can be split in four cases.

Case 1, $\mathrm{h}^0(F(-D)) = 0$ and $\mathrm{h}^0(H(-D)) = 0$.:
Since $\mathrm{H}^0(G(-D))$ sits between two zeros, it must also be zero.

Case 2, $\mathrm{h}^1(F(-D)) = 0$ and $\mathrm{h}^1(H(-D)) = 0$.:
Since $\mathrm{H}^1(G(-D))$ sits between two zeros, it must also be zero.

Case 3, $\mathrm{h}^0(F(-D)) = 0$ **and** $\mathrm{h}^1(H(-D)) = 0$**.:**
If δ_D is injective, then $\mathrm{h}^0(G(-D)) = 0$. Otherwise, if δ_D is surjective, then $\mathrm{h}^1(G(-D)) = 0$.

Case 4, $\mathrm{h}^1(F(-D)) = 0$ **and** $\mathrm{h}^0(H(-D)) = 0$**.:**
If $\mathrm{h}^0(F(-D)) > 0$ and $\mathrm{h}^1(H(-D)) > 0$, then d satisfies Eq. (4.3), so condition (a) is violated. If either of these is zero, we fall back to one of the first three cases.

Finally, G is nonspecial since $\mathrm{H}^1(G)$ sits between $\mathrm{H}^1(F) = 0$ and $\mathrm{H}^1(H) = 0$ in sequence (4.2) for $D = 0$. This proves that G satisfies interpolation, so the converse implication is complete. □

REMARK 4.15. The forward direction of Theorem 4.14 holds without the irreducibility hypothesis on C, that is, if G satisfies interpolation, then statements (a) and (b) are true.

To construct a counterexample for the converse, consider the curve C obtained by gluing two rational components C_1 and C_2 in a single point. Let F be the line bundle obtained by gluing $\mathcal{O}_{C_1}$ and $\mathcal{O}_{C_2}(2)$, and let H be the line bundle obtained by gluing $\mathcal{O}_{C_1}(2)$ and $\mathcal{O}_{C_1}$. Next, we will take $G = F \oplus H$. Both F and H are nonspecial line bundles, so they satisfy interpolation. Condition (a) is automatically satisfied since $\mathrm{h}^0(F) = \mathrm{h}^0(H)$ by the symmetry between F and H. Since G is the direct sum of F and H, it follows that all boundary maps δ_D are zero. To show they are of maximal rank, we need to know that either the source $\mathrm{h}^0(H(-D))$ or the target $\mathrm{h}^1(F(-D))$ is zero. Again, this is true by the symmetry between F and H and the fact both of them satisfy interpolation. Finally, to see that G does not satisfy interpolation note that there exists no degree 3 divisor D such that $\mathrm{h}^0(G(-D)) = 0$ or $\mathrm{h}^1(G(-D)) = 0$.

Specializing F to a line bundle yields the following useful result.

COROLLARY 4.16. *Let F, G, and H be as above, and F is a nonspecial line bundle. If H satisfies interpolation, then G satisfies interpolation if and only if*

(a) $\mathrm{rank}(H)(\mathrm{h}^0(F) - 1) \leq \mathrm{h}^0(H)$, *and*
(b) *for every $d \geq 1$ and a general divisor D of degree d, the boundary map δ_D has maximal rank.*

PROOF. Other than simplifying the inequality in condition (a), this result follows by noting that a line bundle satisfies interpolation if and only if it is nonspecial (Theorem 4.7). □

Let us return to the short exact sequence of vector bundles (4.1). Note that we have the following diagram with exact rows:

$$\begin{array}{ccccccccc} 0 & \longrightarrow & F & \longrightarrow & G & \longrightarrow & H & \longrightarrow & 0 \\ & & \downarrow & & \downarrow & & \| & & \\ 0 & \longrightarrow & F(np) & \longrightarrow & G(np)[np \to F] & \longrightarrow & H & \longrightarrow & 0 \end{array} \tag{4.4}$$

The following proposition (when combined with Proposition 4.12 to remove the positive twist) gives the promised result on modifications along line subbundles without any linear generality assumption.

PROPOSITION 4.17. *Consider diagram (4.4). If*

(a) *F, G, and H satisfy interpolation,*
(b) *F is a line bundle,*
(c) *the point $p \in C_{\mathrm{sm}}$ is general, and*
(d) $\operatorname{rank}(H)(\mathrm{h}^0(F) + n - 1) \leq \mathrm{h}^0(H)$,

then $G(np)[np \to F]$ satisfies interpolation.

PROOF. Both $F(np)$ and H satisfy interpolation (for the first, we apply Theorem 4.11), and

$$\operatorname{rank}(H)(\mathrm{h}^0(F(np)) - 1) = \operatorname{rank}(H)(\mathrm{h}^0(F) + n - 1) \leq \mathrm{h}^0(H).$$

To apply Theorem 4.16 and conclude that $G(np)[np \to F]$ satisfies interpolation, we need to verify that the connecting homomorphism $\delta'_D \colon \mathrm{H}^0(H(-D)) \to \mathrm{H}^1(F(-D+np))$ has maximal rank for D a general divisor of degree d and every $d \geq 1$. On the other hand, we can present δ'_D as a composition using the connecting homomorphism $\delta_D \colon \mathrm{H}^0(H(-D)) \to \mathrm{H}^1(F(-D))$ which has maximal rank (Theorem 4.16).

$$\begin{array}{ccc} \mathrm{H}^0(H(-D)) & \xrightarrow{\delta_D} & H^1(F(-D)) \\ \Big\| & & \Big\downarrow \pi \\ \mathrm{H}^0(H(-D)) & \xrightarrow{\delta'_D} & \mathrm{H}^1(F(-D+np)) \end{array}$$

Since the cokernel of $F(-D) \to F(-D+np)$ is supported at p, it has no higher cohomology, so the morphism $\pi \colon \mathrm{H}^1(F(-D)) \to \mathrm{H}^1(F(-D+np))$ is surjective. If δ_D is surjective, then δ'_D is automatically surjective.

The case when δ_D is injective requires a little more work. Note that the image V of δ_D is independent of the point p. Therefore, it suffices to show that the restriction of π to an arbitrary fixed subspace V has maximal rank.

Set $L = K_C \otimes F(-D)^\vee$, where K_C is the dualizing line bundle (which exists since C is lci). The dual problem asks whether the image of the natural inclusion $\mathrm{H}^0(L(-np)) \to \mathrm{H}^0(L)$ intersects an arbitrary fixed space $V \subset \mathrm{H}^0(L)$ transversely. Since the inclusion has codimension n, by shrinking or enlarging V, it suffices to answer this question when $\dim V = n$. In turn, this is equivalent to the nonvanishing of the Wronskian associated to V [**2**]. (Note that this final step requires our assumption that K has characteristic 0.) □

REMARK 4.18. It is natural to ask whether Theorem 4.17 holds if the rank of F is greater than 1. As presented, the proof does not go through if $\operatorname{rank} F \geq 2$. One of the major obstacles is that the image of $\delta_D \colon \mathrm{H}^0(H(-D)) \to \mathrm{H}^1(F(-D))$ may be contained in $\mathrm{H}^1(F'(-D))$ for some proper subbundle $F' \subset F$.

CHAPTER 5

Elementary modifications of normal bundles

We defined modifications for varieties of arbitrary dimension in Chapter 2, and later provided curve-specific results in Chapter 3. We plan to apply these ideas by using modifications of normal bundles of curves along some very specific subbundles. The aim of the present chapter is to introduce these subbundles and explain their properties.

Let us start with a specific example. Consider a curve $C \subset \mathbb{P}^r$ and a point $p \in \mathbb{P}^r$. We would like to construct a line subbundle $N_{C \to p} \subset N_C$ whose fibers consist of normal directions "pointing to p". To be more specific, choose a smooth point $q \in C_{\mathrm{sm}}$ whose tangent line $[T_q C] \subset \mathbb{P}^r$ does not pass through p. The fiber $N_{C \to p}|_q \subset N_C|_q$ corresponds to the place spanned by p and $[T_q C]$. Note that we made several assumptions about C, p, and q, so the bundle $N_{C \to p}$ may not be defined over the entire curve C. In what follows, we attempt to relax some of these hypotheses while carrying the construction more generally for families of curves. It is also possible to replace the point p by a linear space $\Lambda \subset \mathbb{P}^r$ of arbitrary dimension.

After providing a simple example of what the goal of this section is, we are ready to explain our constructions in full generality. Fix an ambient projective space $\mathbb{P}^r$, and let $\mathcal{C} \subset \mathbb{P}^r \times B$ and $\Lambda \subset \mathbb{P}^r \times B$ respectively be flat families (over B) of curves and linear spaces. The projections on the second factor B will be denoted by $\pi_{\mathcal{C}} \colon \mathcal{C} \to B$ and $\pi_\Lambda \colon \Lambda \to B$. Given a point $b \in B$, we will use $\mathcal{C}_b = \pi_{\mathcal{C}}^{-1}(b)$ and $\Lambda_b = \pi_\Lambda^{-1}(b)$ to denote the curve and linear space over b. For simplicity, we will also assume the base B is reduced and connected.

We define the open set $U_{\mathcal{C},\Lambda} \subset \mathcal{C}$ consisting of all points $p \in \mathcal{C}$ such that $[T_p \mathcal{C}_{\pi_{\mathcal{C}}(p)}] \subset \mathbb{P}^r$, the projective realization of the tangent space to the curve containing p at the same point, does not meet the corresponding linear space $\Lambda_{\pi_{\mathcal{C}}(p)}$. Note that if $p \in \mathcal{C}$ is a smooth point of the fiber $\mathcal{C}_{\pi_{\mathcal{C}}(p)}$, then $[T_p \mathcal{C}_{\pi_{\mathcal{C}}(p)}] \subset \mathbb{P}^r$ is a line. A node point yields a 2-plane $[T_p \mathcal{C}_{\pi_{\mathcal{C}}(p)}]$, while other singularities may lead to even higher dimensional linear spaces. In the cases of interest, all curve singularities will be nodes.

Consider $\varepsilon \colon X = \mathrm{Bl}_\Lambda(\mathbb{P}^r \times B) \to \mathbb{P}^r \times B$ and the projection $\pi_X \colon X \to B$. Let $E = \varepsilon^{-1}(\Lambda)$ denote the exceptional divisor. Since blowing up along a flat subscheme commutes with base change, the fiber $X_b = \pi_X^{-1}(b)$ can be identified with an individual blowup $\mathrm{Bl}_{\Lambda_b} \mathbb{P}^r$. Note that $p \in [T_p \mathcal{C}_{\pi_{\mathcal{C}}(p)}]$ for all $p \in \mathcal{C}$, so $U_{\mathcal{C},\Lambda} \subset \mathcal{C} \setminus \Lambda$. In particular, the embedding $\mathcal{C} \subset \mathbb{P}^r \times B$ lifts to an embedding $U_{\mathcal{C},\Lambda} \to X$. (More generally, there is a lift to $\mathcal{C} \setminus \operatorname{Sing} \pi_{\mathcal{C}}$ but we will not need this fact.)

After constructing the blowup X, it is natural to consider the fiber-by-fiber quotient of $\mathbb{P}^r$ by Λ. If the fibers of Λ have dimension $k < r$, we define

$$Y = \{(P, b) \mid \Lambda_b \subset P\} \subset \mathbb{G}(k, r) \times B$$

with projection to the second factor $\pi_Y : Y \to B$. Similarly to the typical quotient construction, there is a morphism $f : X \to Y$ whose fibers are projective spaces of dimension $k + 1$. Furthermore f descends to a rational morphism $\mathbb{P}^r \times B \dashrightarrow Y$. We have constructed the following commutative diagram over B.

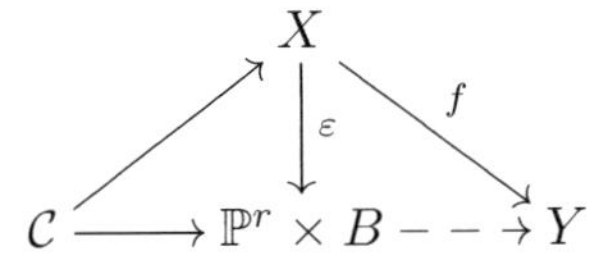

There is a morphism of normal sheaves $N_{\mathcal{C}/X} \to N_{\mathcal{C}/Y}$. If we restrict it to the open $U_{\mathcal{C},\Lambda}$, this becomes a morphism of vector bundles, and the kernel is locally free because the tangent spaces to $\mathcal{C}$ do not meet Λ. We will denote this kernel by $N_{\mathcal{C}\to\Lambda}$. It is important to stress this is a vector bundle only over the open $U_{\mathcal{C},\Lambda}$. Note that $N_{\mathcal{C}\to\Lambda} \subset N_{\mathcal{C}/X}|_{U_{\mathcal{C},\Lambda}}$ by definition. On the other hand, there is a natural isomorphism $N_{\mathcal{C}/X}|_{U_{\mathcal{C},\Lambda}} \cong N_{\mathcal{C}/\mathbb{P}^r \times B}|_{U_{\mathcal{C},\Lambda}}$, so we can treat $N_{\mathcal{C}\to\Lambda}$ as a subbundle of the original normal bundle $N_{\mathcal{C}/\mathbb{P}^r \times B}$.

REMARK 5.1. If the base B is a point, we can extend $N_{\mathcal{C}\to\Lambda}$ to the entire curve $\mathcal{C}$ as long as $U_{\mathcal{C},\Lambda} \neq \emptyset$ contains the singular locus of $\mathcal{C}$. After observing that $N_{C\to\Lambda} \subset N_{\mathcal{C}/\mathbb{P}^r}|_{U_{\mathcal{C},\Lambda}}$, this follows immediately from the curve-to-projective extension theorem.

The widespread use of the bundles $N_{\mathcal{C}\to\Lambda}$ makes it economical to update our modification notation. First, when dealing with a family of normal bundles we will write $[D \to \Lambda]$ instead of $[D \to N_{\mathcal{C}\to\Lambda}]$. Second, if $Z \subset \mathbb{P}^r \times B$ is a family of subvarieties whose fiber-by-fiber spans form a flat family $\Lambda_Z \subset \mathbb{P}^r \times B$, we will also write $[D \to Z]$ instead of $[D \to \Lambda_Z]$.

CHAPTER 6

Examples of the bundles $N_{C\to\Lambda}$

In this chapter, we calculate two important examples of the bundles $N_{C\to\Lambda}$ appearing in the previous chapter. For this, it will be helpful to recall the notion of an Euler vector field.

DEFINITION 6.1. Let $p \in \mathbb{P}V$ and $H \subset \mathbb{P}V$ be a point and a hyperplane, which are complementary. This gives rise to a direct sum decomposition $V \simeq \langle p\rangle \oplus \langle H\rangle$. There is a natural $\mathbb{C}^\times$-action on $\mathbb{P}V$, which is induced by the action of $\lambda \in \mathbb{C}^\times$ on V given by

$$\lambda(x+y) = \lambda x + y \quad \text{for} \quad x \in \langle p\rangle,\ y \in \langle H\rangle.$$

We then define the *Euler vector field* corresponding to p and H as the differential (at $\lambda = 1 \in \mathbb{C}^\times$) of the above action. By inspection, the Euler vector field vanishes at p and along H, but nowhere else.

PROPOSITION 6.2. *Let $\Lambda \simeq \mathbb{P}^{k-1} \subset \mathbb{P}^r$ be a linear subspace with $k \leq r-1$, such that no tangent line of C meets Λ. (So in particular, C does not meet Λ.) Then*

$$N_{C\to\Lambda} \simeq \mathcal{O}_C(1)^k.$$

PROOF. Let $p_1, p_2, \ldots, p_k \in \Lambda$ be k points in linear general position. This gives us a direct sum decomposition

$$N_{C\to\Lambda} \simeq \bigoplus_{j=1}^{k} N_{C\to p_j}.$$

Consequently, we are reduced to the case when $\Lambda = p_1 = p$ is a single point.

In this case, write H for a hyperplane complementary to p. Then the Euler vector field corresponding to p and H gives rise to a section of $N_{C\to p}$ which vanishes precisely along the intersection of C with H. □

PROPOSITION 6.3. *Let $p \in C$ be a point on C, and $k \leq r-1$ be an integer. Suppose that p is not a generalized flex point, and that no other tangent line to C besides T_pC meets $\overline{kp}$. Then there is an isomorphism*

$$\frac{N_{C\to kp}}{N_{C\to(k-1)p}} \simeq \mathcal{O}_C(1)(2p).$$

PROOF. Pick some point $q \in \overline{kp} \smallsetminus \overline{(k-1)p}$. Note that we have a natural map

$$N_{C\to q} \longrightarrow \frac{N_{C\to kp}}{N_{C\to(k-1)p}},$$

which is an isomorphism away from p. Since $N_{C\to q} \simeq \mathcal{O}_C(1)$, we thus have

$$\frac{N_{C\to kp}}{N_{C\to(k-1)p}} \simeq \mathcal{O}_C(1)(np)$$

for some n; it remains to show $n = 2$. For this, we note that the above implies

$$\chi(N_{C\to kp}) - \chi(N_{C\to(k-1)p}) = \chi\left(\frac{N_{C\to kp}}{N_{C\to(k-1)p}}\right) = d - g + 1 + n.$$

In particular, we see that it is sufficient to prove

$$\chi(N_{C\to kp}) = k(d - g + 3).$$

For this, we choose a local coordinate t on C, and an affine patch $p = (0, 0, \ldots) \in \mathbb{A}^r \subset \mathbb{P}^r$, so that in an analytic neighborhood of the origin, C is the locus of points of the form

$$C(t) = (t, t^2 + f_2, t^3 + f_3, \ldots, t^{k+1} + f_{k+1}, f_{k+2}, \ldots),$$

where the $f_i = f_i(t) = O(t^{k+2})$ for all i are holomorphic. (Such a presentation exists because p is not a generalized flex point of C.) Define $q_1, \ldots, q_{k-1}$ via

$$q_i = (\underbrace{0, \ldots, 0}_{i \text{ times}}, 1, 0, 0, \ldots).$$

The Euler vector fields at $p, q_1, \ldots, q_{k-1}$ then define an injective map of sheaves

$$\mathcal{O}_C(1)^k \longrightarrow N_{C\to kp},$$

with cokernel supported at p. Since $\chi(\mathcal{O}_C(1)^k) = k(d - g + 1)$, it remains to show that the cokernel has Euler characteristic $2k$. Equivalently, since the cokernel is supported at p, we want to show that the vectors

$$C(t) = C(t) - p, \quad C(t) - q_1, \quad C(t) - q_2, \quad \ldots, \quad C(t) - q_{k-1}, \quad C'(t)$$

are linearly independent to order exactly $2k$ in t. Or more explicitly, that the $(k + 1) \times r$ matrix

$$\begin{matrix}
t & t^2 + f_2 & \cdots & t^{k-1} + f_{k-1} & t^k + f_k & t^{k+1} + f_{k+1} & f_{k+2} & \cdots \\
-1 + t & t^2 + f_2 & \cdots & t^{k-1} + f_{k-1} & t^k + f_k & t^{k+1} + f_{k+1} & f_{k+2} & \cdots \\
t & -1 + t^2 + f_2 & \cdots & t^{k-1} + f_{k-1} & t^k + f_k & t^{k+1} + f_{k+1} & f_{k+2} & \cdots \\
\vdots & \vdots & \ddots & \vdots & \vdots & \vdots & \vdots & \\
t & t^2 + f_2 & \cdots & -1 + t^{k-1} + f_{k-1} & t^k + f_k & t^{k+1} + f_{k+1} & f_{k+2} & \cdots \\
1 & 2t + f_2' & \cdots & (k-1)t^{k-2} + f_{k-1}' & kt^{k-1} + f_k' & (k+1)t^k + f_{k+1}' & f_{k+2}' & \cdots
\end{matrix}$$

has the property that the minimum power of t dividing the determinants of all $(k + 1) \times (k + 1)$ minors is exactly t^{2k}.

To show this, we may first subtract the first row from rows $2, 3, \ldots, (k - 1)$, to replace the above matrix with:

$$\begin{matrix}
t & t^2 + f_2 & \cdots & t^{k-1} + f_{k-1} & t^k + f_k & t^{k+1} + f_{k+1} & f_{k+2} & \cdots \\
-1 & 0 & \cdots & 0 & 0 & 0 & 0 & \cdots \\
0 & -1 & \cdots & 0 & 0 & 0 & 0 & \cdots \\
\vdots & \vdots & \ddots & \vdots & \vdots & \vdots & \vdots & \\
0 & 0 & \cdots & -1 & 0 & 0 & 0 & \cdots \\
1 & 2t + f_2' & \cdots & (k-1)t^{k-2} + f_{k-1}' & kt^{k-1} + f_k' & (k+1)t^k + f_{k+1}' & f_{k+2}' & \cdots
\end{matrix}$$

Next, we expand along rows $2, 3, \ldots, (k - 1)$, which reduces our problem to showing that the $2 \times (r - k + 1)$ matrix

$$\begin{matrix}
t^k + f_k & t^{k+1} + f_{k+1} & f_{k+2} & f_{k+3} & \cdots \\
kt^{k-1} + f_k' & (k+1)t^k + f_{k+1}' & f_{k+2}' & f_{k+3}' & \cdots
\end{matrix}$$

has the property that the minimum power of t dividing the determinants of all 2×2 minors is exactly t^{2k}. Since this matrix in particular has the form

$$\begin{matrix} O(t^k) & O(t^{k+1}) & O(t^{k+1}) & O(t^{k+1}) & \dots \\ O(t^{k-1}) & O(t^k) & O(t^k) & O(t^k) & \dots \end{matrix}$$

we conclude that the determinant of every 2×2 minor is divisible by t^{2k}. In particular, it suffices to find a particular 2×2 minor whose determinant is not divisible by t^{2k+1}. But we can easily calculate

$$\begin{vmatrix} t^k + f_k & t^{k+1} + f_{k+1} \\ kt^{k-1} + f'_k & (k+1)t^k + f'_{k+1} \end{vmatrix} \equiv t^{2k} \bmod t^{2k+1}. \qquad \square$$

COROLLARY 6.4. *If in addition C is rational, then*

$$N_{C\to kp} \simeq \mathcal{O}_C(1)(2p)^k.$$

PROOF. We argue by induction on k. The base case follows from Corollary 6.3. For the inductive step, Proposition 6.3 gives an exact sequence of vector bundles

$$0 \longrightarrow N_{C\to(k-1)p} \simeq \mathcal{O}_C(1)(2p)^{k-1}$$

$$\simeq \mathcal{O}_{\mathbb{P}^1}(d+2)^{k-1} \longrightarrow N_{C\to kp} \longrightarrow \mathcal{O}_C(1)(2p) \simeq \mathcal{O}_{\mathbb{P}^1}(d+2) \longrightarrow 0.$$

But every exact sequence of vector bundles of the form

$$0 \longrightarrow \mathcal{O}_{\mathbb{P}^1}(n)^a \longrightarrow \bullet \longrightarrow \mathcal{O}_{\mathbb{P}^1}(n)^b \longrightarrow 0$$

on $\mathbb{P}^1$ splits. $\square$

CHAPTER 7

Interpolation and specialization

In order to prove our main theorem, we will argue via degeneration. In this chapter, we set up the basic results necessary for such an argument.

PROPOSITION 7.1. *Given a flat family of curves $\mathcal{C} \to B$, and a vector bundle $\mathcal{E}$ on $\mathcal{C}$, the locus of $b \in B$ for which the pullback $\mathcal{E}_b$ of $\mathcal{E}$ to $\mathcal{C}_b$ satisfies interpolation is open.*

PROOF. The bundle $\mathcal{E}_b$ satisfies interpolation if and only if $\mathrm{H}^0(\mathcal{E}_b(-D)) = 0$ or $\mathrm{H}^1(\mathcal{E}_b(-D)) = 0$, for D a general effective divisor. But the vanishing of a cohomology group is an open condition, hence the result.

For a more careful proof, see Theorem 5.8 of [**1**]. □

We next show that certain constructions for producing reducible curves yield curves which correspond to a point in the component of the Hilbert scheme that we care about.

LEMMA 7.2. *If $C \subseteq \mathbb{P}^r$ is a curve with $\mathrm{H}^1(\mathcal{O}_C(1)) = 0$, then $\mathrm{H}^1(N_C) = 0$.*

PROOF. We begin by considering the Euler sequence

$$0 \longrightarrow \mathcal{O}_C \longrightarrow \mathcal{O}_C(1)^{r+1} \longrightarrow T_{\mathbb{P}^r}|_C \longrightarrow 0,$$

which gives a long exact sequence

$$\cdots \longrightarrow \mathrm{H}^1(\mathcal{O}_C(1))^{r+1} = 0 \longrightarrow \mathrm{H}^1(T_{\mathbb{P}^r}|_C) \longrightarrow \mathrm{H}^2(\mathcal{O}_C) = 0 \longrightarrow \cdots .$$

In particular, we conclude that $\mathrm{H}^1(T_{\mathbb{P}^r}|_C) = 0$. But now from the exact sequence

$$0 \longrightarrow T_C \longrightarrow T_{\mathbb{P}^r}|_C \longrightarrow N_C \longrightarrow 0,$$

we obtain a long exact sequence

$$\cdots \longrightarrow \mathrm{H}^1(T_{\mathbb{P}^r}|_C) = 0 \longrightarrow \mathrm{H}^1(N_C) \longrightarrow \mathrm{H}^2(T_C) = 0 \longrightarrow \cdots .$$

Consequently, $\mathrm{H}^1(N_C) = 0$ as desired. □

DEFINITION 7.3. Write $\mathcal{H}_{d,g,r}$ for the Hilbert scheme of subschemes of $\mathbb{P}^r$ with Hilbert polynomial $P(x) = dx + 1 - g$.

DEFINITION 7.4. We say $C \subseteq \mathbb{P}^r$ is *nonspecial* if it is smooth, irreducible, and $\mathrm{H}^1(\mathcal{O}_C(1)) = 0$. We say C is *limit nonspecial* if C lies in the closure (in the Hilbert scheme $\mathcal{H}_{d,g,r}$) of the locus of nonspecial curves, and satisfies $\mathrm{H}^1(N_C) = 0$.

REMARK 7.5. From Lemma 7.2, we see that every nonspecial curve satisfies $\mathrm{H}^1(N_C) = 0$, and is therefore limit nonspecial.

The set of nondegenerate limit nonspecial curves is parameterized by an open subset of a component of $\mathcal{H}_{d,g,r}$. If $\mathcal{L}$ is a nonspecial line bundle, then $\dim H^0(\mathcal{L}) = d+1-g$. In particular, if $d < g+r$, there are no nonspecial curves (and thus no limit nonspecial curves).

For $d \geq g+r$, the moduli space of nonspecial curves naturally maps to the open subset U of the Picard bundle parameterizing $\{(C, \mathcal{L}) : \mathcal{L} \text{ is nonspecial}\}$. Write $\mathcal{E}$ for the natural vector bundle on U whose fiber over $(C, \mathcal{L})$ gives $H^0(\mathcal{L})$. Then the moduli space of nonspecial curves can be described as $\mathbb{V}_{r+1}(\mathcal{E})$, where $\mathbb{V}_{r+1}$ denotes the variety parameterizing $(r+1)$-frames up to scaling. In particular, since the Picard bundle (and therefore U) is irreducible, the moduli space of nonspecial curves (and thus of limit nonspecial curves) is also irreducible.

DEFINITION 7.6. We say subschemes Y and Z of X meet *quasi-transversely* at $x \in Y \cap Z$ if

$$T_x Y \oplus T_x Z \to T_x X$$

is of maximal rank (i.e. either injective or surjective).

PROPOSITION 7.7 (Hartshorne-Hirschowitz). *Let $C \subset \mathbb{P}^r$ be a curve with $\mathrm{H}^1(N_C) = 0$, and L a line meeting C quasi-transversely at one or two points. Then $C \cup L$ lies in the closure of the locus of smooth irreducible curves and satisfies $\mathrm{H}^1(N_{C\cup L}) = 0$.*

PROOF. See Theorem 4.1 and Corollary 4.2 of [**6**] (which is stated for $r = 3$; but the proof given there generalizes trivially to r arbitrary). □

COROLLARY 7.8. *Let $C \subset \mathbb{P}^r$ be a limit nonspecial curve, and L a line meeting C quasi-transversely at one or two points in C_{sm}. Then $C \cup L$ is limit nonspecial.*

PROOF. By Lemma 7.2 and Proposition 7.7, we know that $C \cup L$ lies in the closure of the locus of smooth irreducible curves and satisfies $\mathrm{H}^1(N_{C\cup L}) = 0$. It thus remains to show that $C \cup L$ lies in the closure of the locus of nonspecial curves.

Since C is a limit nonspecial curve, and we may deform L along with C, we may suppose by semicontinuity that C is itself nonspecial; it thus remains to show that for C nonspecial, $\mathrm{H}^1(\mathcal{O}_{C\cup L}(1)) = 0$. For this, consider the exact sequence of sheaves

$$0 \longrightarrow \mathcal{O}_L(1)(-C \cap L) \longrightarrow \mathcal{O}_{C\cup L}(1) \longrightarrow \mathcal{O}_C(1) \longrightarrow 0,$$

which gives rise to the long exact sequence

$$\cdots \longrightarrow \mathrm{H}^1(\mathcal{O}_L(1)(-C\cap L)) \longrightarrow \mathrm{H}^1(\mathcal{O}_{C\cup L}(1)) \longrightarrow \mathrm{H}^1(\mathcal{O}_C(1)) = 0 \longrightarrow \cdots.$$

To complete the proof, it suffices to show $H^1(\mathcal{O}_L(1)(-C \cap L)) = 0$. But this is clear because $\mathcal{O}_L(1)(-C\cap L) = \mathcal{O}_{\mathbb{P}^1}(1 - \#(C \cap L))$ is either $\mathcal{O}_{\mathbb{P}^1}$ or $\mathcal{O}_{\mathbb{P}^1}(-1)$. □

CHAPTER 8

Reducible curves and their normal bundles

In this chapter, we give several tools for relating interpolation of the normal bundle of a reducible curve to interpolation for the normal bundles of its components. More specifically, we focus on reducible curves of the form $C \cup L$, where L is a line, which we assume for the rest of this section meets C quasi-transversely. Then we can reduce interpolation for the normal bundle $N_{C\cup L}$ to interpolation for a modification N'_C of the normal bundle N_C as follows.

(1) First we find the space of sections $\mathrm{H}^0(N_{C\cup L}|_L(-D))$ which vanish along some divisor D supported on L. Since L is a line, this can be done via explicit computation.

(2) Then we find which sections of $\mathrm{H}^0(N_{C\cup L}|_C)$ match up with the above space of sections along the intersection $C \cap L$. In many cases, this turns out to be expressible in terms of a modification N'_C.

We begin by proving a general proposition that makes precise under what hypotheses the above method is applicable.

PROPOSITION 8.1. *Let $\mathcal{E}$ be a vector bundle on a reducible curve $X \cup Y$, and D be an effective divisor on X disjoint from $X \cap Y$. Assume that*

$$\mathrm{H}^0(\mathcal{E}|_X(-D - X \cap Y)) = 0.$$

Let

$$\mathrm{ev}_X \colon \mathrm{H}^0(\mathcal{E}|_X) \longrightarrow \mathrm{H}^0(\mathcal{E}|_{X\cap Y})$$
$$\mathrm{ev}_Y \colon \mathrm{H}^0(\mathcal{E}|_Y) \longrightarrow \mathrm{H}^0(\mathcal{E}|_{X\cap Y})$$

denote the natural evaluation morphisms. Then $\mathcal{E}$ satisfies interpolation provided that

$$V = \mathrm{ev}_Y^{-1}(\mathrm{ev}_X(\mathrm{H}^0(\mathcal{E}|_X(-D)))) \subseteq \mathrm{H}^0(\mathcal{E}|_Y)$$

satisfies interpolation and has dimension

$$\chi(\mathcal{E}|_Y) + \chi(\mathcal{E}|_X(-D - X \cap Y)).$$

PROOF. We first note that since $\mathrm{H}^0(\mathcal{E}|_X(-D - X \cap Y)) = 0$, the map ev_X is injective when restricted to $\mathrm{H}^0(\mathcal{E}|_X(-D))$. Thus, restriction to Y defines an isomorphism $\mathrm{H}^0(\mathcal{E}(-D)) \simeq V$. Consequently,

$$\mathrm{h}^0(\mathcal{E}(-D)) = \chi(\mathcal{E}|_Y) + \chi(\mathcal{E}|_X(-D - X \cap Y)) = \chi(\mathcal{E}(-D)) \quad \Rightarrow \quad \mathrm{h}^1(\mathcal{E}(-D)) = 0.$$

Since $\mathrm{H}^0(\mathcal{E}(-D)) \simeq V$ satisfies interpolation by assumption, we conclude that $\mathcal{E}(-D)$ — and hence $\mathcal{E}$ — satisfies interpolation. □

We now specialize to the case of $C \cup L$ as in the previous section; we seek to relate interpolation for $N_{C\cup L}$ to interpolation for a modification of N_C, along a divisor supported at $C \cap L$. Since this will depend only on the local behavior of the normal bundles at the nodes $C \cap L$, we can in fact work in slightly greater generality: Suppose that $N'_{C\cup L}$ is a bundle on $C \cup L$, equipped with an isomorphism to $N_{C\cup L}$ over an open set of $C \cup L$ containing the entire line L; in particular, containing a neighborhood U of $C \cap L$ in C.

DEFINITION 8.2. Write N'_C for the bundle obtained from $N'_{C\cup L}|_{C \smallsetminus (C\cap L)}$, glued along $U \smallsetminus (C \cap L)$ via our given isomorphism to the bundle $N_C|_{U\cap C}$. For example, when $N'_{C\cup L} = N_{C\cup L}$, then $N'_C = N_C$.

Now suppose that $u \in C \cap L$ is a point of intersection. Let $v \in L$ and $w \in T_u C$ be points distinct from u.

PROPOSITION 8.3 (Hartshorne-Hirschowitz). *Let V be a neighborhood of u in $C \cap L$, disjoint from the other points of intersection (if any). Then, we have isomorphisms*

$$N'_{C\cup L}|_{C\cap V} \cong N'_C(u)[u \to v]|_{C\cap V} \quad \text{and} \quad N'_{C\cup L}|_{L\cap V} \cong N_L(u)[u \to w]|_{L\cap V},$$

extending the obvious isomorphisms on $V \smallsetminus \{u\}$.

PROOF. The above in the special case of $N'_{C\cup L} = N_{C\cup L}$ follows from Corollary 3.2 of [**6**] (which is stated for $r = 3$, but the proof given applies for r arbitrary). The general case follows from the special case by passing to the neighborhood $U \cap V$ of u. □

From the above, the subbundles $N_{C\to v} \subseteq N_C$ and $N_{L\to w} \subseteq N_L$ give rise to subbundles $N'_{C\to v}(u)|_V \subseteq N'_{C\cup L}|_{C\cap V}$ and $N_{L\to w}(u)|_V \subseteq N'_{C\cup L}|_{L\cap V}$. The key to analyzing these normal bundles is the following result.

LEMMA 8.4. *The fibers of $N'_{C\to v}(u)|_V$ and $N_{L\to w}(u)|_V$ at u coincide.*

PROOF. Consider the cone S parameterizing pairs of points (x, y) with $x \in C$ and y on the line joining x to v. The inclusion $C \hookrightarrow \mathbb{P}^r$ then factors as $C \hookrightarrow S \to \mathbb{P}^r$. (The map $C \hookrightarrow S$ is defined via $x \mapsto (x, x)$; the map $S \to \mathbb{P}^r$ is defined via $(x, y) \mapsto y$.)

Then S is a smooth surface in the neighborhood $C \smallsetminus \{v\}$ of u; so shrinking V if necessary, we may suppose S is smooth along V. Further shrinking V, we may suppose in addition that $U \subseteq V$, so $N'_{C\to v} = N_{C\to v}$.

We now show these fibers coincide by identifying these two bundles with $N_{(C\cup L)/S}|_C$ and $N_{(C\cup L)/S}|_L$. Since two subbundles of a vector bundle coincide if and only if they coincide on a dense open, it suffices to identify these pairs of bundles on $V^\circ = V \smallsetminus \{u\}$. But this is clear because

$$N_{C\to v}(u)|_{V^\circ} = N_{C\to v}|_{V^\circ} = N_{C/S}|_{V^\circ} = N_{(C\cup L)/S}|_{C\cap V^\circ}$$
$$N_{L\to w}(u)|_{V^\circ} = N_{L\to w}|_{V^\circ} = N_{L/S}|_{V^\circ} = N_{(C\cup L)/S}|_{L\cap V^\circ}. \qquad \square$$

LEMMA 8.5. *Suppose L is a 1-secant line to C, meeting C at u; and $p_1, p_2 \in L$ are points distinct from u. Let Λ_1 and Λ_2 be independent linear subspaces of dimensions k_1 and k_2, such that $k_1 + k_2 \leq r - 4$, and $\overline{\Lambda_1 \cdot \Lambda_2}$ is disjoint from $T_u(C \cap L)$. Then*

$$N'_{C\cup L}[p_1 \to \Lambda_1][p_2 \to \Lambda_2]$$

satisfies interpolation provided that

$$N'_C(u)[u \to v][u \to v \cup \Lambda_1 \cup \Lambda_2]$$

satisfies interpolation, where $v \in L$ is any point distinct from u.

PROOF. From Proposition 8.3, we have

$$N'_{C\cup L}|_C \simeq N'_C(u)[u \to v] \quad \Rightarrow \quad N'_{C\cup L}[p_1 \to \Lambda_1][p_2 \to \Lambda_2]|_C \simeq N'_C(u)[u \to v].$$

For T a general linear space of dimension $r - 5 - k_1 - k_2$ (where $\dim \emptyset = -1$ by convention), Proposition 8.3 implies that:

$$\begin{aligned} &N'_{C\cup L}[p_1 \to \Lambda_1][p_2 \to \Lambda_2]|_L \\ &\quad\simeq N_L(u)[u \to w][p_1 \to \Lambda_1][p_2 \to \Lambda_2] \\ &\quad\simeq (N_{L\to w} \oplus N_{L\to\Lambda_1} \oplus N_{L\to\Lambda_2} \oplus N_{L\to T})(u)[u \to w][p_1 \to \Lambda_1][p_2 \to \Lambda_2] \\ &\quad\simeq N_{L\to w}(u - p_1 - p_2) \oplus N_{L\to\Lambda_1}(-p_2) \oplus N_{L\to\Lambda_2}(-p_1) \oplus N_{L\to T}(-p_1 - p_2) \\ &\quad\simeq \mathcal{O}_{\mathbb{P}^1} \oplus \mathcal{O}_{\mathbb{P}^1}^{k_1+1} \oplus \mathcal{O}_{\mathbb{P}^1}^{k_2+1} \oplus \mathcal{O}_{\mathbb{P}^1}(-1)^{r-4-k_1-k_2} \\ &\quad\simeq \mathcal{O}_{\mathbb{P}^1}^{k_1+k_2+3} \oplus \mathcal{O}_{\mathbb{P}^1}(-1)^{r-4-k_1-k_2}. \end{aligned}$$

The positive subbundle $\mathcal{O}_{\mathbb{P}^1}^{k_1+k_2+3}$ here is:

$$N_{L\to w}(u - p_1 - p_2) \oplus N_{L\to\Lambda_1}(-p_2) \oplus N_{L\to\Lambda_2}(-p_1).$$

The above isomorphism also implies:

$$\begin{aligned} &\mathrm{H}^0(N'_{C\cup L}[p_1 \to \Lambda_1][p_2 \to \Lambda_2]|_L(-u)) \\ &\quad\simeq \mathrm{H}^0(\mathcal{O}_{\mathbb{P}^1}(-1)^{k_1+k_2+3} \oplus \mathcal{O}_{\mathbb{P}^1}(-2)^{r-4-k_1-k_2}) = 0 \\ &\chi(N'_{C\cup L}[p_1 \to \Lambda_1][p_2 \to \Lambda_2]|_L(-u)) \\ &\quad= \chi(\mathcal{O}_{\mathbb{P}^1}(-1)^{k_1+k_2+3} \oplus \mathcal{O}_{\mathbb{P}^1}(-2)^{r-4-k_1-k_2}) = k_1 + k_2 + 4 - r. \end{aligned}$$

So applying Proposition 8.1, it suffices to show that the subspace $V \subseteq \mathrm{H}^0(N'_C(u)[u \to v])$ whose fiber at u lies in $\mathrm{H}^0(N_{L\to w}(u - p_1 - p_2)|_u \oplus N_{L\to\Lambda_1}(-p_2)|_u \oplus N_{L\to\Lambda_2}(-p_1)|_u)$ satisfies interpolation and has dimension

$$\chi(N'_C(u)[u \to v]) + k_1 + k_2 + 4 - r = \chi(N'_C(u)[u \to v][u \to v \cup \Lambda_1 \cup \Lambda_2]).$$

But from Lemma 8.4,

$$N_{L\to w}(u - p_1 - p_2)|_u \oplus N_{L\to\Lambda_1}(-p_2)|_u \oplus N_{L\to\Lambda_2}(-p_1)|_u = N'_{C\to v}(u)|_u \oplus N'_{C\to\Lambda_1}|_u \oplus N'_{C\to\Lambda_2}|_u,$$

which implies the above space V is precisely $\mathrm{H}^0(N'_C(u)[u \to v][u \to v\cup\Lambda_1\cup\Lambda_2])$. □

Now we suppose L and C are as above, with $\#(C\cap L) = 2$. Write $C\cap L = \{u, v\}$, and pick general points $x \in T_uC$ and $y \in T_vC$. Let T be a general $(r-4)$-plane in $\mathbb{P}^r$. (We take $T = \emptyset$ if $r = 3$.) We suppose that T_uC does not meet T_vC, so $\{x, y, L\}$ are linearly independent.

DEFINITION 8.6. Let $\mathcal{E} \simeq \mathcal{O}_{\mathbb{P}^1}(1)^k$, and $u, v, z \in \mathbb{P}^1$ be distinct points. Then we obtain an isomorphism

$$\varphi_z\colon \mathrm{H}^0(\mathcal{E}|_u) \xrightarrow{\sim} \mathrm{H}^0(\mathcal{E}|_v),$$

defined by the composition $\mathrm{H}^0(\mathcal{E}|_u) \simeq \mathrm{H}^0(\mathcal{E}(-z)) \simeq \mathrm{H}^0(\mathcal{E}|_v)$.

LEMMA 8.7. *For $z, z' \in \mathbb{P}^1 \smallsetminus \{u, v\}$, we have*

$$\varphi_z = \frac{(z-v)(z'-u)}{(z-u)(z'-v)} \cdot \varphi_{z'}.$$

PROOF. Decomposing $\mathcal{E}$ as a direct sum, we reduce to the case of $k = 1$, in which case the lemma holds by direct computation. □

LEMMA 8.8. *Suppose L is a 2-secant line to C, meeting C at $\{u, v\}$. Then $N'_{C\cup L}$ satisfies interpolation provided that*

$$N'_C(u+v)[u \to v][v \to u][v \to 2u]$$

satisfies interpolation.

PROOF. From Proposition 8.3, we have $N'_{C\cup L}|_C \simeq N'_C(u+v)[u \to v][v \to u]$. Additionally, Proposition 8.3 implies that for $z \in L$ general,

$$\begin{aligned} N'_{C\cup L}|_L(-z) &\simeq N_L(u+v)[u \to x][v \to y](-z) \\ &\simeq \big(N_{L\to x} \oplus N_{L\to y} \oplus N_{L\to T}\big)(u+v)[u \to x][v \to y](-z) \\ &\simeq N_{L\to x}(u-z) \oplus N_{L\to y}(v-z) \oplus N_{L\to T}(-z) \\ &\simeq \mathcal{O}_{\mathbb{P}^1}(2)(-z) \oplus \mathcal{O}_{\mathbb{P}^1}(2)(-z) \oplus \mathcal{O}_{\mathbb{P}^1}(1)(-z)^{r-3} \\ &\simeq \mathcal{O}_{\mathbb{P}^1}(2)(-z)^2 \oplus \mathcal{O}_{\mathbb{P}^1}(1)(-z)^{r-3} \\ &\simeq \mathcal{O}_{\mathbb{P}^1}(1)^2 \oplus \mathcal{O}_{\mathbb{P}^1}^{r-3}. \end{aligned}$$

Note that all isomorphisms except the last one are independent of z. The positive subbundle $\mathcal{O}_{\mathbb{P}^1}(1)^2$ here is canonically:

$$N_{L\to x}(u-z) \oplus N_{L\to y}(v-z);$$

while a (choice of) negative complement is (up to isomorphism independent of z):

$$N_{L\to T}(-z) \simeq \mathcal{O}_{\mathbb{P}^1}(1)(-z)^{r-3}.$$

The above isomorphism also implies:

$$\begin{aligned} \mathrm{H}^0(N'_{C\cup L}|_L(-z-u-v)) &\simeq \mathrm{H}^0(\mathcal{O}_{\mathbb{P}^1}(-1)^2 \oplus \mathcal{O}_{\mathbb{P}^1}(-2)^{r-3}) = 0 \\ \chi(N'_{C\cup L}|_L(-z-u-v)) &= \chi(\mathcal{O}_{\mathbb{P}^1}(-1)^2 \oplus \mathcal{O}_{\mathbb{P}^1}(-2)^{r-3}) = 3-r. \end{aligned}$$

We next examine the map $\mathrm{H}^0(N'_{C\cup L}|_L(-z)) \to \mathrm{H}^0(N'_{C\cup L}|_L(-z)|_{\{u,v\}})$. The above decomposition of $N'_{C\cup L}|_L(-z)$ reduces this problem to understanding the maps

$$\begin{aligned} \mathrm{H}^0(N_{L\to x}(u-z) \oplus N_{L\to y}(v-z)) &\longrightarrow \mathrm{H}^0((N_{L\to x}(u-z) \oplus N_{L\to y}(v-z))|_{\{u,v\}}) \\ \mathrm{H}^0(N_{L\to T}(-z)) &\longrightarrow \mathrm{H}^0(N_{L\to T}(-z)|_{\{u,v\}}). \end{aligned}$$

Since $N_{L\to x}(u-z) \oplus N_{L\to y}(v-z) \simeq \mathcal{O}_{\mathbb{P}^1}(1)^2$, the first map is surjective. The second map is, by construction, the graph of $\varphi_z \colon \mathrm{H}^0(N_{L\to T}|_u) \to \mathrm{H}^0(N_{L\to T}|_v)$. In particular, limiting $z \to v$, Lemma 8.7 implies the image of the second map limits to $\mathrm{H}^0(N_{L\to T}|_u) \times \{0\}$. So applying Proposition 8.1, it suffices to show that the subspace $V \subseteq \mathrm{H}^0(N'_C(u+v)[u \to v][v \to u])$ of sections whose restriction to $\{u, v\}$ lies in

$$\mathrm{H}^0((N_{L\to x}(u-z) \oplus N_{L\to y}(v-z))|_{\{u,v\}}) \oplus (\mathrm{H}^0(N_{L\to T}|_u) \times \{0\}),$$

or equivalently whose restriction to v lies in $\mathrm{H}^0((N_{L\to x}(u-z) \oplus N_{L\to y}(v-z))|_v)$, satisfies interpolation and has dimension

$$\chi(N'_C(u+v)[u \to v][v \to u]) + 3 - r = \chi(N'_C(u+v)[u \to v][v \to u][v \to 2u]).$$

But from Lemma 8.4 and the machinery of Chapter 5, we have

$$(N_{L\to x}(u-z) \oplus N_{L\to y}(v-z))|_v = (N'_{C\to x} \oplus N'_{C\to u}(v))|_v,$$

which gives

$$V = \mathrm{H}^0(N'_C(u+v)[u \to v][v \to u][v \to (u \cup x)]) = \mathrm{H}^0(N'_C(u+v)[u \to v][v \to u][v \to 2u]).$$

□

We now give an alternative condition for $N'_{C \cup L}$ to satisfy interpolation; this requires first introducing some new notation:

DEFINITION 8.9. Let $x \neq y$ be points on C; and X and Y be points of $\mathbb{P}^r$, or sub-line-bundles of N_C. We say a subspace $V \subseteq \mathrm{H}^0(\mathcal{E})$ is an

$$\mathrm{H}^0(\mathcal{E})\langle x \to X : y \to Y\rangle$$

to indicate that

$$\mathrm{H}^0(\mathcal{E}(-x-y)) \subsetneq V \subsetneq \mathrm{H}^0(\mathcal{E}[x \to X][y \to Y]),$$

but that V is neither $\mathrm{H}^0(\mathcal{E}(-x)[y \to Y])$, nor $\mathrm{H}^0(\mathcal{E}(-y)[x \to X])$.

More generally, given points $x_1, y_1, \ldots, x_k, y_k \in C$, and points or sub-line-bundles $X_1, Y_1, \ldots, X_k, Y_k \in \mathbb{P}^r$, we say V is an

$$\mathrm{H}^0(\mathcal{E})\langle x_1 \to X_1 : y_1 \to Y_1\rangle \cdots \langle x_k \to X_k : y_k \to Y_k\rangle$$

if it is a sum (not direct sum) of spaces V_i which are

$$\mathrm{H}^0(\mathcal{E}(-x_1 - y_1 - \cdots - \widehat{x_i} - \widehat{y_i} - \cdots - x_k - y_k))\langle x_i \to X_i : y_i \to Y_i\rangle.$$

In all applications, the $\mathrm{H}^0(E(-x_1 - y_1 - \cdots - \widehat{x_i} - \widehat{y_i} - \cdots - x_k - y_k)[x_i \to X_i]$ $[y_i \to Y_i])$ will be linearly independent relative to $\mathrm{H}^0(E(-x_1 - x_2 - \cdots - x_k - y_k))$.

LEMMA 8.10. *Suppose L is a 2-secant line to C, meeting C at $\{u, v\}$. Write $x \in T_uC$ and $y \in T_vC$ for points in their respective tangent lines, distinct from u and v. Suppose that T_uC does not meet T_vC. Then $N'_{C \cup L}$ satisfies interpolation provided that every*

$$\mathrm{H}^0\left(N'_C(2u+2v)[u \to v][v \to u]\right)\langle u \to v : v \to x\rangle\langle v \to u : u \to y\rangle$$

satisfies interpolation, and

$$\mathrm{H}^1(N'_C[u \to v][v \to u]) = 0.$$

PROOF. From Proposition 8.3, we have $N'_{C \cup L}|_C \simeq N'_C(u+v)[u \to v][v \to u]$. Additionally, Proposition 8.3 implies that for $z, w \in L$ general,

$$\begin{aligned} N'_{C \cup L}|_L(-z-w) &\simeq N_L(u+v)[u \to x][v \to y](-z-w) \\ &\simeq N_{L \to x}(u-z-w) \oplus N_{L \to y}(v-z-w) \oplus N_{L \to T}(-z-w) \\ &\simeq \mathcal{O}_{\mathbb{P}^1}^2 \oplus \mathcal{O}_{\mathbb{P}^1}(-1)^{r-3}. \end{aligned}$$

The above isomorphism also implies:

$$\begin{aligned} \mathrm{H}^0(N'_{C \cup L}|_L(-z-w-u-v)) &\simeq \mathrm{H}^0(\mathcal{O}_{\mathbb{P}^1}(-2)^2 \oplus \mathcal{O}_{\mathbb{P}^1}(-3)^{r-3}) = 0 \\ \chi(N'_{C \cup L}|_L(-z-w-u-v)) &= \chi(\mathcal{O}_{\mathbb{P}^1}(-2)^2 \oplus \mathcal{O}_{\mathbb{P}^1}(-3)^{r-3}) = 4 - 2r. \end{aligned}$$

We next examine the map $\mathrm{H}^0(N'_{C \cup L}|_L(-z-w)) \to \mathrm{H}^0(N'_{C \cup L}|_L(-z-w)|_{\{u,v\}})$. The above decomposition of $N'_{C \cup L}|_L(-z-w)$ reduces this problem to understanding the maps

$$\begin{aligned} \mathrm{H}^0(N_{L\to x}(u-z-w)) &\longrightarrow \mathrm{H}^0(N_{L\to x}(u-z-w)|_{\{u,v\}}) \\ \mathrm{H}^0(N_{L\to y}(v-z-w)) &\longrightarrow \mathrm{H}^0(N_{L\to y}(v-z-w)|_{\{u,v\}}) \\ \mathrm{H}^0(N_{L\to T}(-z-w)) &\longrightarrow \mathrm{H}^0(N_{L\to T}(-z-w)|_{\{u,v\}}). \end{aligned}$$

Since $\mathrm{H}^0(N_{L\to T}(-z-w)) = \mathrm{H}^0(\mathcal{O}_{\mathbb{P}^1}(-1)^{r-3}) = 0$, the last map is zero. The first two maps are, by construction, the graphs of the isomorphisms φ_w:

$$\begin{aligned} \mathrm{H}^0(N_{L\to x}(u-z-w)|_u) &\longrightarrow \mathrm{H}^0(N_{L\to x}(u-z-w)|_v) \\ \mathrm{H}^0(N_{L\to y}(v-z-w)|_u) &\longrightarrow \mathrm{H}^0(N_{L\to y}(v-z-w)|_v). \end{aligned}$$

In particular, they are subsets $W_u \subset \mathrm{H}^0(N'_{L\to x}(u-z-w)|_u)\times \mathrm{H}^0(N'_{L\to x}(u-z-w)|_v)$, respectively $W_v \subset \mathrm{H}^0(N'_{L\to y}(v-z-w)|_u) \times \mathrm{H}^0(N'_{L\to y}(v-z-w)|_v)$, of dimension 1, which are not contained in either factor.

By Proposition 8.1, it suffices to show that the subspace $V \subseteq \mathrm{H}^0(N'_C(u+v)[u\to v][v\to u])$ of sections whose restriction to $\{u,v\}$ lies in $W_u \oplus W_v$ satisfies interpolation and has dimension

$$\chi(N'_C(u+v)[u\to v][v\to u]) + 4 - 2r = \chi(N'_C[u\to v][v\to u]) + 2.$$

We note that by Lemma 8.4, together with the machinery of Chapter 5, the subsets W_u and W_v can equally well be described as subsets:

$$\begin{aligned} W_u &\subset \mathrm{H}^0(N'_{C\to v}(v)|_u) \times \mathrm{H}^0(N'_{C\to x}|_v) \\ W_v &\subset \mathrm{H}^0(N'_{C\to y}|_u) \times \mathrm{H}^0(N'_{C\to u}(v)|_v), \end{aligned}$$

which are of dimension 1 and not contained in either factor.

For the dimension statement, we note that since $\mathrm{H}^1(N'_C[u\to v][v\to u]) = 0$, the map $\underline{\mathrm{res}}$ of restriction to $\{u,v\}$ is surjective. Consequently,

$$\begin{aligned} \dim V &= \dim \mathrm{Ker}(\underline{\mathrm{res}}) + 2 \\ &= \dim \mathrm{H}^0(N'_C[u\to v][v\to u]) + 2 \\ &= \chi(N'_C[u\to v][v\to u]) + 2, \end{aligned}$$

as desired. For the interpolation statement, it suffices to show that V is an

$$\mathrm{H}^0\left(N'_C(2u+2v)[u\to v][v\to u]\right)\langle u\to v : v\to x\rangle\langle v\to u : u\to y\rangle.$$

Write V_u for the subspace of $\mathrm{H}^0(N'_C(u+v)[u\to x][v\to y])$ consisting of sections whose restriction to $\{u,v\}$ lies in W_u, and define similarly V_v. Again, since restriction to $\{u,v\}$ is surjective, $V = V_u + V_v$. It is therefore sufficient (by symmetry) to note that V_u is an

$$\mathrm{H}^0\left(N'_C(u+v)[u\to v][v\to u]\right)\langle u\to v : v\to x\rangle. \qquad \square$$

LEMMA 8.11. *Let $L \subset \mathbb{P}^r$ be a line, and $w, s, t \in \mathbb{P}^r$ be three distinct collinear points, lying on a line which does not meet L. Then for $u, q, p \in L$, the bundle*

$$N_{L\to w\cup s}(u-q)[u\to w][q\to s][p\to t] \simeq \mathcal{O}_{\mathbb{P}^1} \oplus \mathcal{O}_{\mathbb{P}^1}(-1).$$

Moreover, writing $P \simeq \mathcal{O}_{\mathbb{P}^1}$ for the positive subbundle, the fiber $P|_u$ limits, as $q \to u$, to the fiber

$$N_{L\to s}(-q-p)|_u \subseteq N_L(u-q)[u\to w][q\to s][p\to t].$$

PROOF. As $N_{L\to s}$ and $N_{L\to t}$, viewed as subbundles of $N_{L\to w\cup s}$, have distinct fibers at p, this holds after modification at u and q. As $N_{L\to s}(-q)$ is a subbundle of $N_{L\to w\cup s}(u-q)[u\to w][q\to s]$, it follows that $N_{L\to s}(-q-p)$ is a subbundle of $N_{L\to w\cup s}(u)[u\to w][q\to s][p\to t]$. But from Proposition 3.3, we have

$$\chi(N_{L\to s}(-q-p)) = 0$$
$$\chi(N_{L\to w\cup s}(u)[u\to w][q\to s][p\to t]) = 4+2-1-1-1=1.$$

Now by the classification of vector bundles on $\mathbb{P}^1$, is is clear that any rank 2 vector bundle on $\mathbb{P}^1$, which has Euler characteristic 1 and a subbundle of Euler characteristic 0, is necessarily $\mathcal{O}_{\mathbb{P}^1}\oplus\mathcal{O}_{\mathbb{P}^1}(-1)$.

Our next problem is to calculate the behavior of the fiber $P|_u$ as $q\to u$. To do this, we choose isomorphisms $N_{L\to w}\simeq N_{L\to s}\simeq \mathcal{O}_{\mathbb{P}^1}(1)$ such that $N_{L\to t}$ is the diagonal. Here, we identify $\mathcal{O}_{\mathbb{P}^1}(1)$ with the bundle of functions with a pole allowed at ∞. We then act by an automorphism of $\mathbb{P}^1$ preserving ∞ to send u and p to 0 and 1 respectively. We write

$$N_{L\to w\cup s}(u-q)[u\to w][q\to s][p\to t] = (N_{L\to w}(u-2q)\oplus N_{L\to s}(-q))[p\to t].$$

In terms of a local coordinate z on $\mathbb{P}^1$, sections of $N_{L\to w}(u)\oplus N_{L\to s}$ are then expressions of the form

$$\left(\frac{a}{z}+c+dz, b+ez\right).$$

Here, $P|_u$ is identified with the lowest-order terms $[a:b]\in\mathbb{P}(N_{L\to w}(u)|_u\oplus N_{L\to s}|_u)$. To be a section of $(N_{L\to w}(u-2q)\oplus N_{L\to s}(-q))[p\to t]$, we require:

$$\left(\frac{a}{z}+c+dz\right)\Big|_{z=q} = \frac{d}{dz}\left(\frac{a}{z}+c+dz\right)\Big|_{z=q} = 0$$
$$(b+ez)|_{z=q}=0$$
$$\left(\frac{a}{z}+c+dz\right)\Big|_{z=1} = (b+ez)|_{z=1}.$$

This is a system of linear equations in a,b,c,d,e; eliminating c,d,e via elementary linear algebra gives

$$a(1-q)+bq=0.$$

In particular, as $q\to u=0$, the subspace $P|_u$ limits to $[a:b]=[0:1]$. Or in other words, the fiber $P|_u$ limits to the fiber $N_{L\to s}|_u\simeq N_{L\to s}(-q-p)|_u\subseteq N_{L\to w\cup s}(u-q)[u\to w][q\to s][p\to t]$. □

LEMMA 8.12. *Suppose L is a 1-secant line to C, meeting C at u; and $p,q\in L$ and $w\in T_uC$ are points distinct from u. Let Λ be a linear space of dimension $r-3$, and s a point. Suppose the subspaces Λ, s, and w are disjoint from L, and that their projections from L are in linear general position. Then*

$$N'_{C\cup L}[q\to s][p\to\Lambda]$$

satisfies interpolation, in the limit $q\to u$, provided that

$$N'_C[u\to p\cup s]$$

satisfies interpolation.

PROOF. First we note that the bundle we wish to prove satisfies interpolation and the bundle we assume satisfies interpolation, both depend only upon the projection of Λ from L. Consequently, since the projection of Λ from L meets the projection of $\overline{ws}$ from L at a point, we may suppose Λ meets $\overline{ws}$ at a point t. Let $\Lambda' \subset \Lambda$ be a codimension 1 subspace, disjoint from t (which we take to be the empty set if $r = 3$).

Now, from Proposition 8.3, we have

$$N'_{C\cup L}|_C \simeq N'_C(u)[u \to p] \quad \Rightarrow \quad N'_{C\cup L}[q \to s][p \to \Lambda]|_C \simeq N'_C(u)[u \to p].$$

Additionally (using Lemma 8.11),

$$\begin{aligned} &N'_{C\cup L}[q \to s][p \to \Lambda]|_L(-q) \\ &\quad\simeq N_L(u)[u \to w][q \to s][p \to \Lambda](-q) \\ &\quad\simeq N_L(u)[u \to w][q \to s][p \to t \cup \Lambda'](-q) \\ &\quad\simeq (N_{L\to w\cup s} \oplus N_{L\to\Lambda'})(u)[u \to w][q \to s][p \to t \cup \Lambda'](-q) \\ &\quad\simeq N_{L\to w\cup s}(u-q)[u \to w][q \to s][p \to t] \oplus N_{L\to\Lambda'}(-2q) \\ &\quad\simeq \mathcal{O}_{\mathbb{P}^1} \oplus \mathcal{O}_{\mathbb{P}^1}(-1) \oplus \mathcal{O}_{\mathbb{P}^1}(-1)^{r-3} \\ &\quad\simeq \mathcal{O}_{\mathbb{P}^1} \oplus \mathcal{O}_{\mathbb{P}^1}(-1)^{r-2}. \end{aligned}$$

The positive subbundle $\mathcal{O}_{\mathbb{P}^1}$ here is:

$$\text{the positive subbundle } P \text{ of } N_{L\to w\cup s}(u-q)[u \to w][q \to s][p \to t].$$

The above isomorphism also implies:

$$\begin{aligned} \mathrm{H}^0(N'_{C\cup L}[q \to s][p \to \Lambda]|_L(-q-u)) &\simeq \mathrm{H}^0(\mathcal{O}_{\mathbb{P}^1}(-1) \oplus \mathcal{O}_{\mathbb{P}^1}(-2)^{r-3}) = 0 \\ \chi(N'_{C\cup L}[q \to s][p \to \Lambda]|_L(-q-u)) &= \chi(\mathcal{O}_{\mathbb{P}^1}(-1) \oplus \mathcal{O}_{\mathbb{P}^1}(-2)^{r-2}) = 2 - r. \end{aligned}$$

So applying Proposition 8.1, it suffices to show that the subspace $V \subseteq \mathrm{H}^0(N'_C(u)[u \to p])$ whose fiber at u lies in $\mathrm{H}^0(P|_u)$ satisfies interpolation and has dimension

$$\chi(N'_C(u)[u \to p]) + 2 - r = \chi(N'_C[u \to p \cup s]).$$

Again from Lemma 8.11, the subspace $V \subseteq \mathrm{H}^0(N'_C(u)[u \to p])$ thus limits to the space of sections whose fiber at u lies in $\mathrm{H}^0(N_{L\to s}(-p-q)|_u)$, or equivalently (by the machinery of Chapter 5) whose fiber at u lies in $\mathrm{H}^0(N'_{C\to s}|_u)$, which we recognize as the space of sections $\mathrm{H}^0(N'_C(u)[u \to p][u \to s])$. □

CHAPTER 9

A stronger inductive hypothesis

In this chapter, we explain a generalization of our interpolation problem; this generalization will allow us to make an inductive argument in the following chapters.

DEFINITION 9.1. Consider a curve C, equipped with a collection of general points in C_{sm}:

- One marked point p;
- For any triple $(i, j; k)$ of integers in the set

$$(1,1;1) \quad (2,0;1) \quad (1,0;2) \quad (1,1;0) \quad (1,0;1) \quad (2,0;0) \quad (0,0;2) \quad (1,0;0) \quad (0,0;1),$$

  a set of n_{ij}^k points $q_{ij}^k(1), q_{ij}^k(2), \ldots, q_{ij}^k(n_{ij}^k)$. We call these points $(i, j; k)$-*points*, and we write q_{ij}^k for the divisor $q_{ij}^k(1) + \cdots + q_{ij}^k(n_{ij}^k)$.

Here and throughout, we require

$$\sum_{i,j,k} k n_{ij}^k < r - 1.$$

In addition, we require $r \geq 2$; and if $r = 2$, we require

$$\sum_{i,j,k} j n_{ij}^k = 0.$$

Then we define the *modified normal bundle*

$$\begin{aligned} N_C' &= N_C\big((i + j - 1)q_{ij}^k\big)[p \to k q_{ij}^k][i q_{ij}^k \to p][j q_{ij}^k \to 2p] \\ &= N_C\left(\sum_{i,j,k}(i + j - 1)q_{ij}^k\right)\left[p \to \sum_{i,j,k} k q_{ij}^k\right]\left[\sum_{i,j,k} i q_{ij}^k \to p\right]\left[\sum_{i,j,k} j q_{ij}^k \to 2p\right]. \end{aligned}$$

In general, when we write an expression with indices i, j, k in a twist or modification of a vector bundle, the reader should sum over i, j, k, as in the above example.

REMARK 9.2. Note that for every allowed $(i, j; k)$, we have $i \geq j \geq 0$ and $k \geq 0$. Moreover, when $(i, j; k)$ is allowed, then both $(i, j; 0)$ and $(j, 0; k)$ are either allowed or equal to $(0, 0; 0)$. (And in particular, combining these two statements, so is $(j, 0; 0)$.)

Note that when every $n_{ij}^k = 0$, we have $N_C' = N_C(-p)$; hence, interpolation for N_C' will imply interpolation for N_C.

DEFINITION 9.3. We say that a quadruple $\big(d,\ g,\ r,\ n\colon (i, j; k) \mapsto n_{ij}^k\big)$ is *good* if the modified normal bundle of a general curve of degree d and genus g in $\mathbb{P}^r$, with n_{ij}^k general marked points of type $(i, j; k)$, satisfies interpolation.

LEMMA 9.4. *We have*

$$\chi(N'_C) = (r+1)d - (r-3)g - 2 - \sum_{i,j,k} (r-1-i-2j-k)n^k_{ij}.$$

PROOF. Since $\chi(N_C) = (r+1)d-(r-3)(g-1)$, the result follows from counting the changes in the normal bundle at respectively p, and at all $(i,j;k)$-points (c.f. Proposition 3.3):

$$\chi(N'_C) = \chi(N_C) - \left(r - 1 - \sum_{i,j,k} kn^k_{ij}\right) - \sum_{i,j,k} (r-1-i-2j)n^k_{ij}. \qquad \square$$

LEMMA 9.5. *The sub-line-bundle $N'_{C\to p}$ of N'_C consisting of sections which point towards p is nonspecial and has Euler characteristic given by*

$$d - g + 2 + \sum_{i,j,k} (i+j-1)n^k_{ij}.$$

PROOF. The bundle $N_{C\to p}$ (without modification) is isomorphic to $\mathcal{O}_C(1)(2p)$ by Theorem 6.3. Consequently, $N'_{C\to p}$ is isomorphic to

$$N'_{C\to p} \simeq N_{C\to p}(-p)\big((i+j-1)q^k_{ij}\big) \simeq \mathcal{O}_C(1)(p)\big((i+j-1)q^k_{ij}\big).$$

By inspection, $N'_{C\to p}$ is a general line bundle of the given Euler characteristic. It thus remains to show that the given Euler characteristic is positive. But

$$d-g+2+\sum_{i,j,k}(i+j-1)n^k_{ij} \geq (g+r)-g+2-\sum_{i,j,k} kn^k_{ij} \geq r+2-(r-2) \geq 0. \ \square$$

In particular, since by Theorem 4.14 part a, the bundle N'_C can only satisfy interpolation when

$$(r-1)\cdot\chi(N'_{C\to p}) - (r-2) \leq \chi(N'_C),$$

the previous two lemmas imply that a *necessary* condition for N'_C to satisfy interpolation is

$$\sum_{i,j,k} \big((r-2)i + (r-3)j - k\big)\cdot n^k_{ij} \leq 2d + 2g - r - 2. \tag{9.1}$$

Our goal for the rest of the paper will be to establish a partial converse: That subject to certain conditions (the most important of which is (9.1)) — which are satisfied in particular when every $n^k_{ij} = 0$ — the general modified normal bundle N'_C satisfies interpolation. To do this, we must first know that the property of N'_C satisfying interpolation is open; this is made precise by the following crucial proposition.

PROPOSITION 9.6. *Let $\mathcal{C} \to B$ be a family of curves, $p\colon B \to \mathcal{C}$ a section, and q^k_{ij} be effective divisors on $\mathcal{C}$ which are flat over B of relative degree n^k_{ij}. Suppose that, for every $b \in B$:*

1. *The divisor $2p(b) + \sum_{i,j,k} kq^k_{ij}(b)$ is nondegenerate;*
2. *For any $x \in q^k_{ij}(b)$, the tangent lines to $\mathcal{C}(b)$ at x and $p(b)$ are disjoint.*

Then the locus of $b \in B$ so that the modified normal bundle for $(\mathcal{C}(b), p(b), q^k_{ij}(b))$ satisfies interpolation is open.

PROOF. Define

$$\Lambda = \overline{\sum_{i,j,k} k q_{ij}^k} \quad \text{and} \quad P = \overline{2p}.$$

Our first assumption implies that $p = p(B)$ is contained in $U_{\mathcal{C},\Lambda}$. Similarly, our second assumption implies that the q_{ij}^k are contained in $U_{\mathcal{C},P}$ (and consequently in $U_{\mathcal{C},p}$).

We now wish to construct a vector bundle $N'_{\mathcal{C}}$ on the total space $\mathcal{C}$ whose restriction to each fiber $\mathcal{C}(b)$ is $N'_{\mathcal{C}(b)}$. For this, we can appeal to the results of Section 2: It suffices to check that the modification datum

$$(p, N_{C\to\Lambda}), \quad (q', N_{C\to p}), \quad (q'', N_{C\to P})$$

is tree-like, where

$$q' = \sum_{i,j,k} i n_{ij}^k \quad \text{and} \quad q'' = \sum_{i,j,k} j q_{ij}^k.$$

Since our second condition implies that p does not meet either q' or q'', it suffices to check $\{N_{C\to p}, N_{C\to P}\}$ is tree-like along $q' \cap q''$. But this is clear, since $N_{C\to p} \subset N_{C\to P}$.

The desired result now follows from applying Theorem 7.1 to our bundle $N'_{\mathcal{C}}$. □

CHAPTER 10

Inductive arguments

In this chapter, we give a number of inductive arguments to reduce interpolation for certain modified normal bundles to interpolation for other "simpler" modified normal bundles. We begin by two ways of adding a 2-secant line, which result from respectively limiting u and v to p in Lemma 8.8.

Throughout this chapter, and in the following chapter, we will make use of several such "limiting arguments", all but one of which are straight-forward applications of the machinery developed in Chapter 2. In Theorem 10.2 below, we will spell this out this limiting argument explicitly; subsequently, starting with Theorem 10.5, it will be left to the reader to check that the limiting argument given in Theorem 10.2 applies, mutatis mutandis.

We will also spell out the limiting argument explicitly in Theorem 10.3, as this is the only case where the argument given in Theorem 10.2 (mutatis mutandis) does not apply.

PROPOSITION 10.1. *If a modified normal bundle N'_C satisfies interpolation, then so does a general negative twist*

$$N'_C(-p_1 - p_2 - \cdots - p_n) \quad \text{for} \quad n \leq r + 1 - \sum_{i,j,k} k n_{ij}^k.$$

PROOF. Since N'_C satisfies interpolation, it satisfies Eq. (9.1). By casework, we see that for each allowed $(i, j; k)$,

$$r - 1 - i - 2j - k \leq (r-2)i + (r-3)j + (r-2)k.$$

Combining these facts and applying Theorem 9.4,

$$\begin{aligned}
\chi(N'_C) &= (r+1)d - (r-3)g - 2 - \sum_{i,j,k} (r - 1 - i - 2j - k) n_{ij}^k \\
&\geq (r+1)d - (r-3)g - 2 - \sum_{i,j,k} ((r-2)i + (r-3)j + (r-2)k) n_{ij}^k \\
&= (r+1)d - (r-3)g - 2 \\
&\qquad - \sum_{i,j,k} ((r-2)i + (r-3)j - k) n_{ij}^k - (r-1) \cdot \sum_{i,j,k} k n_{ij}^k \\
&\geq (r+1)d - (r-3)g - 2 - (2d + 2g - r - 2) - (r-1) \cdot \sum_{i,j,k} k n_{ij}^k
\end{aligned}$$

$$= (r-1)(d-g) + r - (r-1) \cdot \sum_{i,j,k} kn_{ij}^k$$

$$\geq (r-1) \cdot r + (r-1) - (r-1) \cdot \sum_{i,j,k} kn_{ij}^k$$

$$= (r-1) \cdot \left(r + 1 - \sum_{i,j,k} kn_{ij}^k \right).$$

Consequently, for $n \leq r + 1 - \sum_{i,j,k} kn_{ij}^k$, the twist $N'_C(-p_1 - p_2 - \cdots - p_n)$ has nonnegative Euler characteristic, which immediately implies the desired conclusion. □

LEMMA 10.2. *Let $g > 0$. Suppose that $(d, g, r; n)$ satisfies (9.1) and*

$$\sum_{i,j,k} kn_{ij}^k < r - 2.$$

Then $(d, g, r; n)$ is good provided that $(d-1, g-1, r; n')$ is good, where

$$(n')_{ij}^k = \begin{cases} n_{ij}^k & \text{if } (i,j;k) \neq (1,1;1); \\ n_{11}^1 + 1 & \text{if } (i,j;k) = (1,1;1). \end{cases}$$

If instead

$$\sum_{i,j,k} kn_{ij}^k = r - 2,$$

then $(d, g, r; n)$ is good provided that $(d-1, g-1, r; n')$ is good, where

$$(n')_{ij}^k = \begin{cases} \sum_\ell n_{ij}^\ell & \text{if } k = 0 \text{ and } (i,j;k) \notin \{(0,0;0),(1,1;0)\}; \\ 1 + \sum_\ell n_{ij}^\ell & \text{if } (i,j;k) = (1,1;0); \\ 0 & \text{else.} \end{cases}$$

PROOF. Degenerate C to $D \cup L$, where L is a 2-secant line, and all marked points specialize to points on D. Applying Lemma 8.8, interpolation for N'_C is reduced to interpolation for

$$N'_D(u+v)[u \to v][v \to u][v \to 2u].$$

The next step is to *limit* $u \to p$, which reduces our problem to interpolation for

$$N'_D(p+v)[p \to v][v \to p][v \to 2p].$$

More precisely, write $\mathcal{D} \to B$ for the constant family $D \times B \to B$ over B, where $B \subset D$ is some open set containing p. What we mean by the above is that there is a vector bundle on $\mathcal{D}$, whose restriction to the fiber $D \times \{u\}$ is, for $u \in B$,

$$N'_D(u+v)[u \to v][v \to u][v \to 2u].$$

For this, we let Λ, P, q', and q'' be as in Theorem 9.6; and write $T = \sum_{i,j,k}(i + j - 1)q_{ij}^k$. By minor abuse of notation, we also write v for the horizontal divisor $v \times B$, and u for the diagonal in $D \times B$. We then appeal to the machinery of Section 2, which constructs our desired bundle

$$N_\mathcal{D}(T)[p \to \Lambda][q' \to p][q'' \to P](u+v)[u \to v][v \to u][v \to 2u],$$

provided that the modification datum

$$(p, N_{\mathcal{D}\to\Lambda}), \quad (q', N_{\mathcal{D}\to p}), \quad (q'', N_{\mathcal{D}\to P}), \quad (u, N_{\mathcal{D}\to v}), \quad (v, N_{\mathcal{D}\to u}), \quad (v, N_{\mathcal{D}\to\overline{2u}})$$

is tree-like. The divisor p does not cross either q' or q''; additionally, since v is general, v does not cross p, q', or q''. Moreover, by shrinking B, we may suppose u does not cross v, q', or q''. It therefore suffices to see that the collections of bundles

$$\{N_{\mathcal{D}\to p}, N_{\mathcal{D}\to P}\}, \quad \{N_{\mathcal{D}\to u}, N_{\mathcal{D}\to \overline{2u}}\}, \quad \{N_{\mathcal{D}\to \Lambda}, N_{\mathcal{D}\to v}\}$$

are tree-like along $q' \cap q''$, v, and $(p,p) \in \mathcal{D} = D \times B$ respectively. But $N_{\mathcal{D}\to p} \subset N_{\mathcal{D}\to P}$ and $N_{\mathcal{D}\to u} \subset N_{\mathcal{D}\to \overline{2u}}$, which takes care of the first two cases. For the last case, the generality of v implies v is not contained in the span of Λ with the tangent line to D at p. Consequently, the fibers $N_{\mathcal{D}\to u}|_{(p,p)}$ and $N_{\mathcal{D}\to \overline{2u}}|_{(p,p)}$ are linearly independent.

Moving on, we collect together the transformations $[p \to \sum kq_{ij}^k(\ell)]$ and $[p \to u]$ into a single transformation $[p \to u + \sum kq_{ij}^k(\ell)](-p)$ via Proposition 2.20.

When $\sum_{i,j,k} kn_{ij}^k < r-2$, we recognize this as another modified normal bundle, with a new point of type $(1,1;1)$ introduced at u, as desired.

Similarly, when $\sum_{i,j,k} kn_{ij}^k < r-2$, we recognize its twist by $-p$ as another modified normal bundle, where all points of type $(i,j;k)$ are changed to type $(i,j;0)$, and a new point of type $(1,1;0)$ is introduced at u. Eliminating the $(0,0;0)$-points (which are just general negative twists) via Theorem 10.1, we reduce to interpolation for a bundle assumed to satisfy interpolation. □

LEMMA 10.3. *Let $g > 0$ and $r > 3$, and suppose that $(d,g,r;n)$ satisfies* (9.1) *and*

$$\sum_{i,j,k} kn_{ij}^k \in \{r-3, r-2\}.$$

Then $(d,g,r;n)$ is good provided that $(d-1,g-1,r;n')$ is good, where for

$$\sum_{i,j,k} kn_{ij}^k = r-3,$$

we have

$$(n')_{ij}^k = \begin{cases} \sum_\ell n_{ij}^\ell & \text{if } k=0 \text{ and } (i,j;k) \neq (0,0;0), \\ 1 & \text{if } (i,j;k) = (1,0;1), \\ 0 & \text{else;} \end{cases}$$

and for

$$\sum_{i,j,k} kn_{ij}^k = r-2,$$

we have

$$(n')_{ij}^k = \begin{cases} \sum_\ell n_{ij}^\ell & \text{if } k=0 \text{ and } (i,j;k) \neq (0,0;0), \\ 1 & \text{if } (i,j;k) = (1,0;2), \\ 0 & \text{else.} \end{cases}$$

PROOF. Again, we degenerate C to $D \cup L$, where L is a 2-secant line, and all marked points specialize to points on D. Twisting the formula in Lemma 8.8 by $-u$, interpolation for N_C' is reduced to interpolation for

$$N_D'(v)[u \to v][v \to u][v \to 2u] \simeq N_D'(v)[u \to v][v \to 2u][v \to u].$$

We now limit $v \to p$, to reduce our problem to interpolation for

$$N_D'(p)[u \to p][p \to u][p \to 2u] \simeq N_D'(p)[u \to p][p \to 2u][p \to u].$$

Since this is the only case in the paper where the limiting argument explained in Theorem 10.2 does not apply (the modification datum in question is not necessarily tree-like), we elaborate further. As in Theorem 10.2, write $\mathcal{D} \to B$ for the constant family $D \times B \to B$ over B, where $B \subset D$ is some open set containing p. Then we want a vector bundle on $\mathcal{D}$, whose restriction to the fiber $D \times \{v\}$ is, for $v \in B$,

$$N'_D(u+v)[u \to v][v \to u][v \to 2u].$$

By minor abuse of notation, we write u for the horizontal divisor $u \times B$, and v for the diagonal in $D \times B$. As u is general, u does not meet p or any q_{ij}^k. Moreover, by shrinking B, we may suppose v does not intersect the tangent line to D at u; by the machinery of Chapter 5, we obtain a subbundle $N_{\mathcal{D} \to v} \subset N_{\mathcal{D}}$. Applying the theory of Chapter 2, this subbundle corresponds to a subbundle $N'_{\mathcal{D} \to v} \subset N'_{\mathcal{D}}$ in a neighborhood of u (where it is tree-like with respect to our modification datum). However, the subbundles $N_{\mathcal{D} \to u} \subset N_{\mathcal{D}}$ and $N_{\mathcal{D} \to \overline{2u}} \subset N_{\mathcal{D}}$ need not be tree-like with respect to our modification datum. To get around this, we invoke the theory of Chapter 3: Since D is a curve, the subbundles $N_{D \to u} \subset N_D$ and $N_{D \to \overline{2u}} \subset N_D$ correspond to subbundles $N'_{D \to u} \subset N'_D$ and $N'_{D \to \overline{2u}} \subset N'_D$. We then let

$$N'_{\mathcal{D} \to u} = \pi^*(N'_{D \to u}) \subset \pi^*(N'_D) \simeq N'_{\mathcal{D}}$$

and

$$N'_{\mathcal{D} \to \overline{2u}} = \pi^*(N'_{D \to \overline{2u}}) \subset \pi^*(N'_D) \simeq N'_{\mathcal{D}},$$

where $\pi \colon \mathcal{D} = D \times B \to D$ is the projection. The machinery of Section 2 then constructs our desired bundle

$$N'_{\mathcal{D}}(u+v)[u \to N'_{\mathcal{D} \to v}][v \to N'_{\mathcal{D} \to u}][v \to N'_{\mathcal{D} \to \overline{2u}}].$$

(This modification datum is tree-like, since u does not cross v, and $N'_{\mathcal{D} \to u} \subset N'_{\mathcal{D} \to \overline{2u}}$.)

Moving on, if $\sum_{i,j,k} kn_{ij}^k = r - 3$, we then collect together the transformations $[p \to kq_{ij}^k]$ and $[p \to 2u]$ (occurring in the right expression) into a negative twist via Proposition 2.20. This yields another modified normal bundle, where all points of type $(i, j; k)$ are changed to type $(i, j; 0)$, and a new point of type $(1, 0; 1)$ is introduced at u. Eliminating all $(0, 0; 0)$-points, we arrive at the desired conclusion.

Similarly, if $\sum_{i,j,k} kn_{ij}^k = r - 2$, we collect together the transformations $[p \to kq_{ij}^k]$ and $[p \to u]$ (occurring in the left expression) into a negative twist via Proposition 2.20. This yields another modified normal bundle, where all points of type $(i, j; k)$ are changed to type $(i, j; 0)$, and a new point of type $(1, 0; 2)$ is introduced at u. Eliminating all $(0, 0; 0)$-points, we arrive at the desired conclusion. □

LEMMA 10.4. *Let $r = 5$ and $g \geq 2$. Write C for a general curve of degree $d - 2$ and genus $g - 2$ in $\mathbb{P}^5$, with markings given by n, and fix general points $q, x, y \in C$. Then $(d, g, 5; n)$ is good provided that*

$$N'_C[q \to x + y][x + y \to q]$$

satisfies interpolation.

In particular, if $n = \mathbf{0}$, then $(d, g, 5; \mathbf{0})$ is good provided that $(d - 2, g - 2, 5; n')$ is good, where

$$(n')_{ij}^k = \begin{cases} 2 & \text{if } (i, j; k) = (1, 0; 1); \\ 0 & \text{otherwise.} \end{cases}$$

PROOF. We degenerate a general curve of degree d and genus g in $\mathbb{P}^5$ to a union $C \cup L \cup M$, where L and M are 2-secant lines to C, and all marked points specialize to points on C. Write $C \cap L = \{x, z\}$ and $C \cap M = \{y, w\}$. By Theorem 8.8, it suffices to show interpolation for

$$N'_C(x+y+z+w)[x \to z][z \to x][z \to 2x][y \to w][w \to y][w \to 2y].$$

Limiting z and w to a common point q reduces the above to interpolation for

$$\begin{aligned} &N'_C(x+y+2q)[x \to q][q \to x][q \to 2x][y \to q][q \to y][q \to 2y] \\ &\simeq N'_C(x+y)[q \to x+y][x+y \to q], \end{aligned}$$

which follows from our assumption that $N'_C[q \to x+y][x+y \to q]$ satisfies interpolation. □

We now give several techniques to reduce from interpolation of modified normal bundles of curves in a give projective space, to interpolation for curves in a projective space of smaller dimension. The basic construction here is to add a line transverse to a hyperplane to a curve contained in that hyperplane. We also explore variants with adding two lines.

LEMMA 10.5. *Suppose that*

$$2d + 2g - 3r + 2 \leq \sum_{i,j,k} ((r-2)i + (r-3)j - k) \cdot n_{ij}^k \leq 2d + 2g - r - 2.$$

If in addition

$$\sum_{i,j,k} k n_{ij}^k < r - 2,$$

then $(d, g, r; n)$ is good provided that $(d-1, g, r-1; n')$ is good, where

$$(n')_{ij}^k = \begin{cases} \sum_\ell n_{\ell i}^k & \text{if } j = 0 \text{ and } (i, j; k) \neq (0, 0; 0); \\ 0 & \text{else.} \end{cases}$$

If instead

$$\sum_{i,j,k} k n_{ij}^k = r - 2,$$

then $(d, g, r; n)$ is good provided that $(d-1, g, r-1; n')$ is good, where

$$(n')_{ij}^k = \begin{cases} \sum_{\ell, m} n_{\ell i}^m & \text{if } j = k = 0 \text{ but } i \neq 0; \\ 0 & \text{else.} \end{cases}$$

PROOF. We degenerate C to $D \cup L$ where $D \subset H$ is a curve contained in a hyperplane, and L is a 1-secant line to D transverse to H. We specialize so $p \in L$ and all other marked points lie on D. Clearly, it suffices to show

$$\mathcal{E} = N'_{D \cup L}(-z)$$

satisfies interpolation, where $z \in L$. From Lemma 8.5 (with $\Lambda_1 = \overline{kq_{ij}^k}$ and $\Lambda_2 = \emptyset$), we conclude it is sufficient to prove interpolation for the bundle

$$\mathcal{E} = N_D\big((i+j-1)q_{ij}^k\big)[iq_{ij}^k \to p][jq_{ij}^k \to 2p](x)[x \to p]\Big[x \to p + \sum kq_{ij}^k\Big]$$

Identifying $\mathcal{O}_D(1)$ with the normal bundle of D in the cone $\overline{pD}$, we obtain a splitting:

$$N_D \simeq N_{D/H} \oplus \mathcal{O}_D(1).$$

This induces a splitting $\mathcal{E} \simeq \mathcal{F} \oplus \mathcal{L}$ with

$$\mathcal{F} = N_{D/H}\big((j-1)q_{ij}^k\big)[jq_{ij}^k \to x][x \to kq_{ij}^k] \quad \text{and} \quad \mathcal{L} = \mathcal{O}_D(1)\big(x + (i+j-1)q_{ij}^k\big).$$

Now we claim $\mathcal{F}$ satisfies interpolation. Indeed, when $\sum kn_{ij}^k < r-2$, then $\mathcal{F}$ is a modified normal bundle of the type assumed to satisfy interpolation. Otherwise, when $\sum kn_{ij}^k = r-2$, then $\mathcal{F}(-x)$ is a modified normal bundle of the type assumed to satisfy interpolation.

Next, $\mathcal{L}$ satisfies interpolation since $\mathcal{O}_D(1)$ satisfies interpolation. So to check $\mathcal{F} \oplus \mathcal{L}$ satisfies interpolation, we just need to check

$$(r-2) \cdot (\chi(\mathcal{L}) - 1) \leq \chi(\mathcal{F}) \leq (r-2) \cdot (\chi(\mathcal{L}) + 1).$$

For this, we first calculate

$$\chi(\mathcal{F}) = r(d-1) - (r-4)g - 2 - \sum_{i,j,k} (r-2-j-k) \cdot n_{ij}^k,$$

$$\begin{aligned}\chi(\mathcal{L}) &= (d-1) - g + 1 + 1 + \sum_{i,j,k} (i+j-1) \cdot n_{ij}^k \\ &= d - g + 1 + \sum_{i,j,k} (i+j-1) \cdot n_{ij}^k.\end{aligned}$$

The condition for $\mathcal{F} \oplus \mathcal{L}$ to satisfy interpolation is then

$$2d + 2g - 3r + 2 \leq \sum_{i,j,k} ((r-2)i + (r-3)j - k) \cdot n_{ij}^k \leq 2d + 2g - r - 2. \qquad \square$$

LEMMA 10.6. *Suppose that* $r > 3$ *and*

$$2d + 2g - 4r + 3 \leq \sum_{i,j,k} ((r-2)i + (r-3)j - k) \cdot n_{ij}^k \leq 2d + 2g - 2r - 1.$$

If in addition

$$\sum_{i,j,k} kn_{ij}^k < r - 3,$$

then $(d, g, r; n)$ *is good provided that* $(d-2, g-1, r-1; n')$ *is good, where*

$$(n')_{ij}^k = \begin{cases} \sum_\ell n_{\ell i}^k & \text{if } j = 0 \text{ and } (i,j;k) \notin \{(0,0;0), (2,0,1)\}; \\ 1 + \sum_\ell n_{\ell i}^k & \text{if } (i,j;k) = (2,0,1); \\ 0 & \text{else.} \end{cases}$$

If instead

$$\sum_{i,j,k} kn_{ij}^k = r - 3,$$

then $(d, g, r; n)$ *is good provided that* $(d-2, g-1, r-1; n')$ *is good, where*

$$(n')_{ij}^k = \begin{cases} \sum_{\ell,m} n_{\ell i}^m & \text{if } j = k = 0 \text{ and } i \notin \{0, 2\}; \\ 1 + \sum_{\ell,m} n_{\ell i}^m & \text{if } j = k = 0 \text{ and } i = 2; \\ 0 & \text{else.} \end{cases}$$

If instead

$$\sum_{i,j,k} kn_{ij}^k = r - 2,$$

then $(d, g, r; n)$ *is good provided that* $(d-2, g-1, r-1; n')$ *is good, where*

$$(n')_{ij}^k = \begin{cases} \sum_{\ell,m} n_{\ell i}^m & \text{if } j = k = 0 \text{ and } i \neq 0; \\ 1 & \text{if } (i, j; k) = (2, 0, 1); \\ 0 & \text{else.} \end{cases}$$

PROOF. We degenerate C to $D \cup L \cup M$ where $D \subset H$ is a curve contained in a hyperplane, and L and M are 1-secant lines to D transverse to H, which meet at some point $q \notin H$. Write $x = L \cap D$ and $s = M \cap D$.

We specialize so $p \in L$ and all other marked points lie on D. Applying Lemma 8.8, and twisting by $-q$, it suffices to show

$$N'_{D\cup L}(s)[q \to s][s \to q][s \to 2q] = N'_{D\cup L}(s)[q \to s][s \to q][s \to p \cup x]$$

satisfies interpolation.

First suppose $\sum kn_{ij}^k < r - 2$. Then limiting $q \to x$, we conclude from Lemma 8.5 (with $\Lambda_1 = s$ and $\Lambda_2 = \overline{kq_{ij}^k}$) that it is sufficient to prove interpolation for the bundle

$$\mathcal{E} = N_D\big((i + j - 1)q_{ij}^k\big)[iq_{ij}^k \to p][jq_{ij}^k \to 2p](s)[s \to x] \\ \cdot [s \to p \cup x](x)[x \to p]\Big[x \to p + s + \sum kq_{ij}^k\Big].$$

Similarly, for $\sum kn_{ij}^k = r - 2$, we conclude by limiting $q \to x$ and applying Lemma 8.12 that it is sufficient to prove interpolation for the bundle

$$\mathcal{E}' = N_D\big((i + j - 1)q_{ij}^k\big)[iq_{ij}^k \to p][jq_{ij}^k \to 2p](s)[s \to x][s \to p \cup x][x \to p \cup s].$$

Identifying $\mathcal{O}_D(1)$ with the normal bundle of D in the cone $\overline{pD}$, we obtain the splitting

$$N_D \simeq N_{D/H} \oplus \mathcal{O}_D(1).$$

This induces splittings $\mathcal{E} \simeq \mathcal{F} \oplus \mathcal{L}$ and $\mathcal{E}' \simeq \mathcal{F}' \oplus \mathcal{L}'$, where:

$$\begin{aligned} \mathcal{F} &= N_{D/H}\big(s + (j-1)q_{ij}^k\big)[jq_{ij}^k \to x][2s \to x]\Big[x \to s + \sum kq_{ij}^k\Big], \\ \mathcal{L} &= \mathcal{O}_D(1)\big(x + (i + j - 1)q_{ij}^k\big), \\ \mathcal{F}' &= N_{D/H}\big(s + (j-1)q_{ij}^k\big)[jq_{ij}^k \to x][2s \to x][x \to s], \\ \mathcal{L}' &= \mathcal{O}_D(1)\big((i + j - 1)q_{ij}^k\big). \end{aligned}$$

Now we claim $\mathcal{F}$, respectively $\mathcal{F}'$, satisfies interpolation. Indeed, when $\sum kn_{ij}^k < r - 3$, then $\mathcal{F}$ is a modified normal bundle of the type assumed to satisfy interpolation. Otherwise, when $\sum n_{ij}^k = r - 3$, then $\mathcal{F}(-x)$ is a modified normal bundle of the type assumed to satisfy interpolation. Finally, when $\sum n_{ij}^k = r - 2$, then $\mathcal{F}'$ is a modified normal bundle of the type assumed to satisfy interpolation.

Next, $\mathcal{L}$, respectively $\mathcal{L}'$, satisfies interpolation since $\mathcal{O}_D(1)$ satisfies interpolation. So to check $\mathcal{F} \oplus \mathcal{L}$, respectively $\mathcal{F}' \oplus \mathcal{L}'$, satisfies interpolation, we just need to check

$$(r-2) \cdot (\chi(\mathcal{L}) - 1) \leq \chi(\mathcal{F}) \leq (r-2) \cdot (\chi(\mathcal{L}) + 1),$$
$$(r-2) \cdot (\chi(\mathcal{L}') - 1) \leq \chi(\mathcal{F}') \leq (r-2) \cdot (\chi(\mathcal{L}') + 1).$$

For this, we first calculate

$$\begin{aligned}
\chi(\mathcal{F}) &= r(d-2) - (r-4)(g-1) - 2 - (r-5) - \sum_{i,j,k} (r-2-j-k) \cdot n_{ij}^k \\
&= r(d-2) - (r-4)g - 1 - \sum_{i,j,k} (r-2-j-k) \cdot n_{ij}^k, \\
\chi(\mathcal{L}) &= (d-2) - (g-1) + 1 + 1 + \sum_{i,j,k} (i+j-1) \cdot n_{ij}^k \\
&= d - g + 1 + \sum_{i,j,k} (i+j-1) \cdot n_{ij}^k, \\
\chi(\mathcal{F}') &= r(d-2) - (r-4)(g-1) - 2 - (r-5) - \sum_{i,j,k} (r-2-j) \cdot n_{ij}^k \\
&= r(d-2) - (r-4)g - 1 - (r-2) - \sum_{i,j,k} (r-2-j-k) \cdot n_{ij}^k, \\
\chi(\mathcal{L}') &= (d-2) - (g-1) + 1 + \sum_{i,j,k} (i+j-1) \cdot n_{ij}^k \\
&= d - g + \sum_{i,j,k} (i+j-1) \cdot n_{ij}^k.
\end{aligned}$$

Substituting this into the above, the condition for either $\mathcal{F} \oplus \mathcal{L}$, or $\mathcal{F}' \oplus \mathcal{L}'$, to satisfy interpolation is then

$$2d + 2g - 4r + 3 \leq \sum_{i,j,k} ((r-2)i + (r-3)j - k) \cdot n_{ij}^k \leq 2d + 2g - 2r - 1. \qquad \square$$

Next, we give an inductive construction based around adding a 1-secant line to C.

LEMMA 10.7. *Let $d > g+r$. If $r = 3$, then assume in addition that $\sum_{i,j,k} j n_{ij}^k = 0$. Suppose that*

$$\sum_{i,j,k} (r - 1 - i - 2j - k) n_{ij}^k \leq (r+1)d - (2r-4)g - 2.$$

Then $(d, g, r; n)$ is good provided that both $(d-1, g, r; n)$ and $(d-1, g, r-1; n')$ are good, where if

$$\sum_{i,j,k} k n_{ij}^k < r - 2,$$

then

$$n' = n;$$

and if instead

$$\sum_{i,j,k} k n_{ij}^k = r - 2,$$

then

$$(n')_{ij}^k = \begin{cases} \sum_\ell n_{ij}^\ell & \text{if } k = 0 \text{ and } (i,j;k) \neq (0,0;0), \\ 0 & \text{else.} \end{cases}$$

PROOF. We degenerate C to $D \cup L$, where D is a nondegenerate curve and L is a 1-secant line to D. We suppose that $s := C \cap L$ is general, and we write q for some other point on L. For q general, projection from q gives a local immersion from D to $\mathbb{P}^{r-1}$.

The image D' under projection is by construction a general curve of degree $d-1$ and genus g in $\mathbb{P}^{r-1}$, and all marked points on D' are general. We now consider

$$\widetilde{N}'_{D'} := N_{D'}\big((i+j-1)q_{ij}^k\big)[p \to kq_{ij}^k][iq_{ij}^k \to p][jq_{ij}^k \to 2p].$$

For for $\sum_{i,j,k} kn_{ij}^k < r-2$, this is a modified normal of the type assumed to satisfy interpolation. Otherwise, for $\sum_{i,j,k} kn_{ij}^k = r-2$, then $\widetilde{N}'_{D'}(-p)$ is a modified normal of the type assumed to satisfy interpolation. Either way, we conclude that $\widetilde{N}'_{D'}$ satisfies interpolation.

Our assumed inequality implies, via Proposition 4.12, that it suffices to prove interpolation for the bundle $N'_C(\Delta_0)$ for any effective divisor $\Delta_0 \subset D \subset C$. From Lemma 8.5 (with $\Lambda_1 = \Lambda_2 = \emptyset$), this in turn reduces to interpolation for

$$N'_D(\Delta_0)(s)[2s \to q].$$

Taking $\Delta_0 = \Delta + s$ for Δ a general effective divisor of large degree, it suffices to prove interpolation for

$$N'_D(\Delta)(2s)[2s \to q] = N'_D(\Delta)(2s)[2s \to N_{D\to q}(-p-q_{ij}^k)].$$

Because the quotient

$$N'_D(\Delta)/N_{D\to q}(\Delta) \simeq \widetilde{N}'_{D'}(\Delta)$$

satisfies interpolation, we can apply Proposition 4.17 to reach the desired conclusion, subject to the inequality

$$(r-2)\cdot\big(\chi(N_{D\to q}(-p-q_{ij}^k)(\Delta))+1\big) \le \chi(N'_D(\Delta)) - \chi(N_{D\to q}(-p-q_{ij}^k)(\Delta)).$$

(For Δ of large degree, $N_{D\to q}(-p-q_{ij}^k)(\Delta)$ will be nonspecial.) This inequality is in turn equivalent to the inequality

$$(r-2)\cdot\big(\chi(N_{D\to q}(-p-q_{ij}^k))+1\big) \le \chi(N'_D) - \chi(N_{D\to q}(-p-q_{ij}^k)).$$

By Proposition 6.2 and Lemma 9.4, we have

$$\chi(N_{D\to q}(-p-q_{ij}^k)) = (d-1) - g + 1 - 1 - \sum_{i,j,k} n_{ij}^k$$

$$\chi(N'_D) = (r+1)(d-1) - (r-3)g - 2 - \sum_{i,j,k}(r-1-i-2j-k)n_{ij}^k.$$

Thus, we just need to check the inequality

$$r+2 \le 2d + 2g + \sum_{i,j,k}(i+2j+k)n_{ij}^k.$$

But this inequality holds since $2d+2g \ge 2(g+r+1)+2g \ge 2r+2 \ge r+2$ by assumption. $\square$

We now give several methods for getting rid of marked points (without changing the degree or genus of our curve).

LEMMA 10.8. *Suppose that* $(d, g, r; n)$ *satisfies* (9.1), *and that*

$$\sum_{i,j,k}(r-1-i-2j-k)\cdot n_{ij}^k \le (r+1)d - (2r-4)g - 2.$$

If there is some integer ℓ with and $n^0_{\ell 0} > 0$, then $(d, g, r; n)$ is good provided that both $(d, g, r; n')$ and $(d-1, g, r-1; n'')$ are good. Here, we define

$$(n')^k_{ij} = \begin{cases} n^k_{ij} - 1 & \text{if } (i, j; k) = (\ell, 0; 0); \\ n^k_{ij} & \text{else.} \end{cases}$$

In addition, if

$$\sum_{i,j,k} k n^k_{ij} < r - 2,$$

then we define

$$(n'')^k_{ij} = \begin{cases} \sum_\ell n^k_{\ell i} & \text{if } j = 0 \text{ and } (i, j; k) \neq (0, 0; 0); \\ 0 & \text{else.} \end{cases}$$

If instead

$$\sum_{i,j,k} k n^k_{ij} = r - 2,$$

then we define

$$(n'')^k_{ij} = \begin{cases} \sum_{\ell,m} n^m_{\ell i} & \text{if } j = k = 0 \text{ but } i \neq 0; \\ 0 & \text{else.} \end{cases}$$

PROOF. Let q be some point of type $(\ell, 0; 0)$. From Proposition 4.12, it is sufficient to prove that $N'_C(q)$ satisfies interpolation, since the given inequality rearranges to $\chi(N'_C) \geq (r-1)g$ (c.f. Lemma 9.4). By assumption, some modification N''_C of type $(d, g, r; n')$ satisfies interpolation. Then we can write

$$N'_C(q) = N''_C(\ell q)[\ell q \to p],$$

Proposition 4.17 implies this satisfies interpolation, as long as N''_C, $N''_{C\to p}$, and $N''_C/N''_{C\to p}$ all satisfy interpolation, and $(r-2)(\chi(N''_{C\to p}) + \ell - 1) \leq \chi(N''_C/N''_{C\to p})$.

We first note that N''_C satisfies interpolation, since $(d, g, r; n')$ is good by assumption; in addition, $N''_{C\to p}$ satisfies interpolation, since it is a nonspecial line bundle by Theorem 9.5. Write $\tilde{C}$ for the proper transform of C in the blowup $\mathrm{Bl}_p \mathbb{P}^r$; let $\bar{C} \subset \mathbb{P}^{r-1}$ denote the projection of C from p. Then using the exact sequence

$$0 \to N_{C\to p}(-p) \to N_C(-p) \simeq N_{\tilde{C}/\mathrm{Bl}_p \mathbb{P}^r} \to N_{\bar{C}/\mathbb{P}^{r-1}} \to 0,$$

we recognize $(N''_C/N''_{C\to p})(-\alpha p)$ as a modified normal bundle, for $\bar{C}$, of type $(d-1, g, r-1; n'')$, where

$$\alpha = \begin{cases} 2 & \text{if } \sum_{i,j,k} k n^k_{ij} < r - 2; \\ 1 & \text{if } \sum_{i,j,k} k n^k_{ij} = r - 2. \end{cases}$$

In particular, our assumption that $(d-1, g, r-1; n'')$ is good implies that

$$(N''_C/N''_{C\to p})(-\alpha p),$$

and thus $N''_C/N''_{C\to p}$, satisfies interpolation. It thus remains only to check

$$(r-2)(\chi(N''_{C\to p}) + \ell - 1) \leq \chi(N''_C/N''_{C\to p}).$$

To do this, we first calculate (using Theorem 9.4 and Theorem 9.5):

$$\begin{aligned}
\chi(N''_{C\to p}) &= d - g + 2 + \sum_{i,j,k}(i+j-1)(n')^k_{ij} \\
&= d - g + 3 - \ell + \sum_{i,j,k}(i+j-1)n^k_{ij} \\
\chi(N''_C/N''_{C\to p}) &= (r-2) + r(d-1) - (r-4)g - 2 - \sum_{i,j,k}(r-2-j-k)(n')^k_{ij} \\
&= rd - (r-4)g + r - 6 - \sum_{i,j,k}(r-2-j-k)n^k_{ij}.
\end{aligned}$$

Substituting the above expressions into our desired inequality reduces it to (9.1), which holds by assumption. □

LEMMA 10.9. *Let $r = 3$, and suppose that $(d, g, r; n)$ satisfies* (9.1)*. If there are integers ℓ and $m \geq 1$ with $n^0_{\ell m} > 0$, then $(d, g, r; n)$ is good provided that $(d, g, r; n')$ is good, where*

$$(n')^k_{ij} = \begin{cases} n^k_{ij} - 1 & \text{if } (i, j; k) = (\ell, m; 0); \\ n^k_{ij} & \text{else.} \end{cases}$$

PROOF. Let q be some point of type $(\ell, m; 0)$. Then since $m \geq 1$, it is sufficient to prove that $N'_C\big(-(m-1)q\big)$ satisfies interpolation. By assumption, some modification N''_C of type $(d, g, r; n')$ satisfies interpolation. Then we can write

$$N'_C\big(-(m-1)q\big) = N''_C(\ell q)[\ell q \to p].$$

Proposition 4.17 implies this satisfies interpolation, as long as N''_C, $N''_{C\to p}$, and $N''_C/N''_{C\to p}$ all satisfy interpolation, and $\chi(N''_{C\to p}) + \ell - 1 \leq \chi(N''_C/N''_{C\to p})$.

We first note that N''_C satisfies interpolation, since $(d, g, r; n')$ is good by assumption. For the remaining conditions, we first note that by Theorem 6.3,

$$\begin{aligned}
N''_{C\to p} &\simeq \mathcal{O}_C(1)(p)\big((i+j-1)(q')^k_{ij}\big) \\
\wedge^2 N''_C &\simeq (\wedge^2 N_C)\big((i+2j-2)(q')^k_{ij}\big)\big((k(n')^k_{ij})\cdot p\big)(-2p) \\
&\simeq K_C(4)\big((i+2j-2)(q')^k_{ij}\big)\big((k(n')^k_{ij})\cdot p\big)(-2p).
\end{aligned}$$

In particular,

$$N''_C/N''_{C\to p} \simeq K_C(3)\big((j-1)(q')^k_{ij}\big)\big(((k(n')^k_{ij}) - 3)\cdot p\big).$$

These expressions imply in particular that $N''_{C\to p}$ and $N''_C/N''_{C\to p}$ are general line bundles on C. By Theorem 9.5, the line bundle $N''_{C\to p}$ is nonspecial; since $\ell \geq 1$, it therefore remains only to check

$$\chi(N''_{C\to p}) + \ell - 1 \leq \chi(N''_C/N''_{C\to p}).$$

For this, we calculate

$$\begin{aligned}
\chi(N''_C/N''_{C\to p}) - \chi(N''_{C\to p}) &= 2g + 2g - 6 - \sum_{i,j,k}(i-k)\cdot(n')^k_{ij} \\
&= 2g + 2g - 6 + \ell - \sum_{i,j,k}(i-k)\cdot n^k_{ij}.
\end{aligned}$$

From (9.1),

$$\sum_{i,j,k} (i-k) \cdot n_{ij}^k \leq 2d + 2g - 5,$$

which implies

$$\chi(N''_C / N''_{C\to p}) - \chi(N''_{C\to p}) \geq 2g + 2g - 6 + \ell - (2d + 2g - 5) = \ell - 1,$$

as desired. □

LEMMA 10.10. *Suppose again that*

$$\sum_{i,j,k} (r - 1 - i - 2j - k) \cdot n_{ij}^k \leq (r+1)d - (2r-4)g - 2.$$

If there is some integer ℓ with and $n_{00}^\ell > 0$, then $(d, g, r; n)$ is good provided that $(d, g, r; n')$ and $(d, g, r; n'')$ are both good, where

$$(n')_{ij}^k = \begin{cases} n_{ij}^k - 1 & \textit{if } (i, j; k) = (0, 0; \ell); \\ n_{ij}^k & \textit{else.} \end{cases}$$

$$(n'')_{ij}^k = \begin{cases} \sum_m n_{ij}^m & \textit{if } (i, j) \neq (0, 0) \textit{ and } k = 0; \\ 0 & \textit{else.} \end{cases}$$

PROOF. Let q be some point of type $(0, 0; \ell)$. Again from Proposition 4.12, it is sufficient to prove that $N'_C(q)$ satisfies interpolation. Write $(q_{ij}^k)^\circ$ for the divisor q_{ij}^k, minus q if $(i, j; k) = (0, 0; \ell)$. Then by assumption, some modification N''_C of type $(d, g, r; n'')$ satisfies interpolation. Then we can write

$$N'_C(q) = N''_C\big(-(q_{ij}^k)^\circ\big)(p)[p \to kq_{ij}^k].$$

Since N''_C satisfies interpolation, we conclude that $N''_C\big(-(q_{ij}^k)^\circ\big)(p)$ does as well. So applying Theorem 4.10, we conclude that N'_C satisfies interpolation. □

CHAPTER 11

Base cases

In this section, we prove interpolation in a number of special cases, which will form the base cases for our inductive argument. These include cases where $r = 2$, for which a general nonspecial line bundle on a general curve C does not define a immersion to $\mathbb{P}^r$. In these cases, we therefore write $N_C = \ker(f^*\Omega_{\mathbb{P}^r} \to \Omega_C)^\vee$ for the normal bundle of the map $f\colon C \to \mathbb{P}^r$ (as opposed to the normal bundle of the image $f(C)$, which coincides when f is an immersion).

LEMMA 11.1. *Suppose that $r = 2$ and $\sum_{i,j,k} jn_{ij}^k = 0$. Then $(d, g, r; n)$ is good.*

PROOF. Our earlier assumption that $\sum_{i,j,k} kn_{ij}^k < r - 1 = 1$ implies, together with our given assumption, that $n_{ij}^k = 0$ unless $j = k = 0$. The modified normal bundle in this case is then

$$N_C' = N_C\big((i-1)q_{ij}^k\big).$$

But from Lemma 7.2, we have $\mathrm{H}^1(N_C) = 0$; consequently, $\mathrm{H}^1(N_C') = 0$. Because N_C is a line bundle, this implies N_C satisfies interpolation. $\square$

LEMMA 11.2. *Suppose that $r = 3$ and $g > 0$, and that*

$$2d + 2g - 9 \leq \sum_{i,j,k} (i-k) \cdot n_{ij}^k \leq 2d + 2g - 7.$$

Then $(d, g, r; n)$ is good.

PROOF. We degenerate C to $D \cup L \cup M$ where $D \subset H$ is a curve contained in a plane, and L and M are 1-secant lines to D transverse to H, which meet at some point $q \notin H$. Write $x = L \cap D$ and $s = M \cap D$.

We specialize so $p \in L$ and all other marked points lie on D. Applying Lemma 8.8, and twisting by $-q$, it suffices to show

$$N_{D\cup L}'(s+q)[q \to s][s \to q][s \to 2q](-q) = N_{D\cup L}'(s)[q \to s][s \to q]$$

satisfies interpolation.

First suppose $\sum kn_{ij}^k = 0$. Then limiting $q \to x$, we conclude from Lemma 8.5 (with $\Lambda_1 = s$ and $\Lambda_2 = \emptyset$) that it is sufficient to prove interpolation for the bundle

$$\mathcal{E}_0 = N_D\Big(\sum (i+j-1)q_{ij}^k\Big)[iq_{ij}^k \to p][jq_{ij}^k \to 2p](s)[s \to x](x)[x \to p][x \to p \cup s].$$

Similarly, suppose $\sum kn_{ij}^k = 1$. Then limiting $q \to x$, we conclude from Lemma 8.12 that it is sufficient to prove interpolation for the bundle

$$\mathcal{E}_1 = N_D\Big(\sum (i+j-1)q_{ij}^k\Big)[iq_{ij}^k \to p][jq_{ij}^k \to 2p](s)[s \to x][x \to p \cup s].$$

Identifying $\mathcal{O}_D(1)$ with its normal bundle in the cone $\overline{pD}$, we obtain a splitting:

$$N_D \simeq N_{D/H} \oplus \mathcal{O}_D(1).$$

This induces splittings $\mathcal{E}_\alpha \simeq \mathcal{F} \oplus \mathcal{L}_\alpha$ for $\alpha = 0, 1$ with

$$\mathcal{F} = N_{D/H}\big(s + (j-1)q_{ij}^k\big)[jq_{ij}^k \to x]$$

and

$$\mathcal{L}_\alpha = \mathcal{O}_D(1)\Big((1-\alpha)x + \sum (i+j-1)q_{ij}^k\Big).$$

Both $\mathcal{F}$ and $\mathcal{L}_\alpha$ satisfy interpolation: for $\mathcal{F}$ this follows from Lemma 11.1, and for $\mathcal{L}_\alpha$ this is immediate from $\mathrm{H}^1(\mathcal{O}_D(1)) = 0$. So to check $\mathcal{E}_\alpha = \mathcal{F} \oplus \mathcal{L}_\alpha$ satisfies interpolation, we just need to check

$$|\chi(\mathcal{L}_\alpha) - \chi(\mathcal{F})| \leq 1.$$

For this, we first calculate

$$\begin{aligned}
\chi(\mathcal{F}) &= 3(d-2) + (g-2) + 1 + \sum (j-1)n_{ij}^k \\
&= 3d + g - 7 + \sum (j-1)n_{ij}^k. \\
\chi(\mathcal{L}_\alpha) &= (d-2) - (g-1) + 1 + 1 - \alpha + \sum_{i,j,k} (i+j-1) \cdot n_{ij}^k \\
&= d - g + 1 - \sum_{i,j,k} k n_{ij}^k + \sum_{i,j,k} (i+j-1) \cdot n_{ij}^k.
\end{aligned}$$

We conclude that

$$\begin{aligned}
&|\chi(\mathcal{L}_\alpha) - \chi(\mathcal{F})| \\
&= \left| d - g + 1 - \sum_{i,j,k} k n_{ij}^k + \sum (i+j-1) \cdot n_{ij}^k - 3d - g + 7 - \sum (j-1) n_{ij}^k \right| \\
&= \left| \sum (i-k) \cdot n_{ij}^k - 2d - 2g + 8 \right| \\
&\leq 1.
\end{aligned}$$

□

LEMMA 11.3. *For $g \geq 1$, the tuple $(5g+1, g, 4g+1; \mathbf{0})$ is good.*

PROOF. We will construct an explicit curve of degree $5g+1$ and genus g in $\mathbb{P}^{4g+1}$ (with no marked points); and check directly that its normal bundle satisfies interpolation.

Let $D \subset \mathbb{P}^{4g+1}$ be a rational normal curve, and let $L_1, L_2, \ldots, L_g$ be a collection of g lines which are 2-secant to D. Then by construction, $D \cup L_1 \cup \cdots \cup L_g$ is a curve of degree $5g+1$ and genus g in $\mathbb{P}^{4g+1}$.

Our task is now to show N_D satisfies interpolation. Write $L_i \cap D = \{x_i, y_i\}$. Then from Corollary 6.4, we obtain

$$N_{D \to 2x_i} \simeq N_{D \to 2y_i} \simeq \mathcal{O}_{\mathbb{P}^1}(4g+3)^2.$$

We have a map of vector bundles

$$\bigoplus_i (N_{D \to 2x_i} \oplus N_{D \to 2y_i}) \to N_D,$$

which is an isomorphism over the generic point. Moreover,

$$\chi\left(\bigoplus_i (N_{D \to 2x_i} \oplus N_{D \to 2y_i})\right) = 4g(4g+4) = \chi(N_D);$$

which implies that in fact the above map yields an isomorphism

$$N_D \simeq \bigoplus_i (N_{D\to 2x_i} \oplus N_{D\to 2y_i}). \tag{11.1}$$

Now from Lemma 8.10, writing $z_i \in T_{x_i}D$ and $w_i \in T_{y_i}D$ for points distinct from x_i and y_i respectively, it suffices to show interpolation for every

$$H^0 N_D(2x_i + 2y_i)[x_i \to y_i][y_i \to x_i])\langle x_i \to y_i : y_i \to z_i\rangle\langle y_i \to x_i : x_i \to w_i\rangle.$$

Happily, each of these transformations respects the direct sum decomposition (11.1): the above space of sections is a direct sum of spaces of sections of each bundle on the RHS of (11.1). By symmetry, it therefore suffices to show interpolation for every

$$\begin{aligned}\mathrm{H}^0(N_{D\to 2x_1} \oplus N_{D\to 2y_1})&(2x_1 + 2y_1 - x_2 - y_2 - \cdots - x_g - y_g)\\ &[x_1 \to y_1][y_1 \to x_1]\langle x_1 \to y_1 : y_1 \to z_1\rangle\langle y_1 \to x_1 : x_1 \to w_1\rangle.\end{aligned} \tag{11.2}$$

Now given a bundle $\mathcal{E} \simeq \mathcal{L} \oplus \mathcal{L}$ with $\mathcal{L}$ a line bundle on a variety X, splittings of $\mathcal{E}$ are in bijection with splittings of any fiber $\mathcal{E}|_x$ (for $x \in X$). In particular, given such a bundle, an inclusion $\iota\colon \mathcal{L} \hookrightarrow \mathcal{E}$, and any vector $v \in \mathcal{E}|_x$ for some $x \in X$ with $v \notin \iota(\mathcal{L}|_x)$, there is a splitting so that ι is inclusion into the first factor and v is an element of the fiber of the second factor. Applying this here, we can choose a splitting

$$N_{D\to 2x_1} \simeq N_{D\to x_1} \oplus N_{D\to x_1}^{\perp} \quad \text{with} \quad N_{D\to x_1} \simeq N_{D\to x_1}^{\perp} \simeq \mathcal{O}_{\mathbb{P}^1}(4g+3),$$

so that

$$N_{D\to x_1}^{\perp}|_{y_1} = N_{D\to z_1}|_{y_1} \subset N_{D\to 2x_1}|_{y_1}.$$

Similarly, we define $N_{D\to y_1}^{\perp}$. Then thanks to Remark 2.2, the space of sections (11.2) splits as a direct sum of two spaces of sections: the space

$$\begin{aligned}\mathrm{H}^0(N_{D\to x_1} \oplus N_{D\to y_1}^{\perp})&(2x_1 + 2y_1 - x_2 - y_2 - \cdots - x_g - y_g)\\ &(-x_1)[y_1 \to x_1](-x_1 - y_1)\langle y_1 \to x_1 : x_1 \to w_1\rangle,\end{aligned}$$

and the space obtained by reversing the roles of x and y above. It thus, by symmetry, suffices to prove interpolation for the above space of sections; we can rewrite this as

$$\begin{aligned}\mathrm{H}^0(N_{D\to x_1} \oplus N_{D\to y_1}^{\perp})&(y_1 - x_2 - y_2 - \cdots - x_g - y_g)\\ &[y_1 \to N_{D\to x_1}]\langle y_1 \to N_{D\to x_1} : x_1 \to N_{D\to y_1}^{\perp}\rangle.\end{aligned}$$

Under the isomorphisms $N_{D\to x_1} \simeq N_{D\to x_1}^{\perp} \simeq \mathcal{O}_{\mathbb{P}^1}(4g+3)$, the above space of sections becomes

$$\mathrm{H}^0(\mathcal{O}_{\mathbb{P}^1}(2g+6) \oplus \mathcal{O}_{\mathbb{P}^1}(2g+5))\langle y_1 \to \mathcal{O}_{\mathbb{P}^1}(2g+6) : x_1 \to \mathcal{O}_{\mathbb{P}^1}(2g+5)\rangle.$$

This is some subspace of sections which is codimension one in

$$\mathrm{H}^0(\mathcal{O}_{\mathbb{P}^1}(2g+6) \oplus \mathcal{O}_{\mathbb{P}^1}(2g+5)),$$

but which by definition in particular does not contain the first factor $\mathrm{H}^0(\mathcal{O}_{\mathbb{P}^1}(2g+6))$. But by direct inspection, every such subspace of sections satisfies interpolation. $\square$

LEMMA 11.4. *The tuple* $(8,3,5;n)$ *is good, where*

$$n_{ij}^k = \begin{cases} 2 & \text{if } (i,j;k) = (1,0;1); \\ 0 & \text{otherwise.} \end{cases}$$

PROOF. Write $q_{10}^1 = \{s,t\}$, and write E for a general curve of degree 6 and genus 1 in $\mathbb{P}^5$. Applying Theorem 10.4, we need to prove interpolation for

$$\mathcal{F} = N_E'[q \to x+y][x+y \to q] = N_E[s+t \to p][p \to s+t][q \to x+y][x+y \to q].$$

Degenerate E to $C \cup L$, where C is a rational normal curve, and L is a 2-secant line. We specialize s and x to L, and all other marked points to C. Write $E \cap L = \{u,v\}$, and let $z \in T_uE$ and $w \in T_vE$ be points distinct from u and v respectively. Then by Theorem 8.3,

$$\mathcal{F}|_L = N_L(u+v)[u \to z][v \to w][s \to p][x \to q] \simeq \mathcal{O}_L^{\oplus 4}.$$

Applying Theorem 8.1 to $\mathcal{F}$, with D a single general point on L, it suffices to show that the space of sections

$$\mathrm{H}^0(\mathcal{F}|_C(-u-v)) = \mathrm{ev}_C^{-1}(\mathrm{ev}_L(\mathrm{H}^0(\mathcal{F}|_L(-D)))) \subseteq \mathrm{H}^0(\mathcal{F}|_C)$$

satisfies interpolation and has dimension

$$\chi(\mathcal{F}|_C) + \chi(\mathcal{F}|_L(-D-x-y)) = \chi(\mathcal{F}|_C(-x-y)).$$

In other words, it remains to prove interpolation for the bundle

$$\mathcal{F}|_C(-u-v) = N_C[t \to p][p \to s+t][q \to x+y][y \to q][u \to v][v \to u].$$

Limiting $x \to p$ and $s \to q$, this reduces to interpolation for

$$N_C[t \to p][p \to q+t][q \to p+y][y \to q][u \to v][v \to u].$$

Further limiting $y \to u$ and $t \to v$, this reduces to interpolation for

$$\begin{aligned} &N_C[v \to p][p \to q+v][q \to p+u][u \to q][u \to v][v \to u] \\ &\simeq N_{C\to p}(-2u-v-p) \oplus N_{C \to q}(-2v-u-q) \\ &\quad \oplus N_{C\to u}(-2u-v-p) \oplus N_{C\to v}(-2v-u-q) \\ &\simeq \mathcal{O}_{\mathbb{P}^1}(3)^{\oplus 4}. \end{aligned}$$

To see the first isomorphism above, we note that there is a natural injection of sheaves from $N_{C\to p} \oplus N_{C\to q} \oplus N_{C\to u} \oplus N_{C\to v}$ to N_C; since they are both vector bundles of the same Euler characteristic, the cokernel must be zero. The final isomorphism to $\mathcal{O}_{\mathbb{P}^1}(3)^{\oplus 4}$ is provided by Theorem 6.3. This completes the proof, since $\mathcal{O}_{\mathbb{P}^1}(3)^{\oplus 4}$ clearly satisfies interpolation. □

CHAPTER 12

Summary of Remainder of Proof of Theorem 1.2

To finish the proof of Theorem 1.2, it remains to show that our collection of inductive statements and base cases from the preceeding two chapters combine to show that every tuple $(d, g, r; \mathbf{0})$ with $d \geq g + r$ and $(d, g, r) \notin \{(5, 2, 3), (6, 2, 4), (7, 2, 5)\}$ is good. This is a purely combinatorial problem, but one which requires a rather involved argument (as hinted by the presence of exactly three exceptions); hence we defer the proof to Appendix A, providing a brief outline here:

(1) We begin by calculating how our various inductive arguments interact with Equation (9.1).
(2) Next, we show that our base cases and inductive arguments imply (9.1) is a sufficient condition for the modified normal bundle of a rational curve to satisfy interpolation.
(3) We then show our base cases and inductive arguments imply (9.1) is sufficient for the modified normal bundle of a space curve to satisfy interpolation — apart from two infinite families, which contain only finitely many members with $n = \mathbf{0}$. Except $(d, g, r) = (5, 2, 3)$, these $n = \mathbf{0}$ cases are also good by ad-hoc application of our base cases and inductive arguments.
(4) Then we show there are finitely many $(d, g, r; n)$ which are not good for $4 \leq r \leq 11$, and use a computer program (c.f. Appendix B) to search over all possible applications of our base cases and inductive arguments, thereby greatly reducing the size of the finite list (and showing in particular that all $(d, g, r; n)$ with $9 \leq r \leq 11$ are good).
(5) Finally, we apply our base cases and inductive arguments to show $(d, g, r; n)$ is good for $r \geq 12$, unless certain inequalities and congruence conditions modulo 5 (which force $n \neq \mathbf{0}$) are satisfied.

CHAPTER 13

The three exceptional cases

In this chapte, we show conversely that if C is a general curve of degree d and genus g in $\mathbb{P}^r$, where

$$(d, g, r) \in \{(5, 2, 3), (6, 2, 4), (7, 2, 5)\}, \tag{13.1}$$

then N_C does not satisfy interpolation. In these cases, we also determine when C passes through general points $p_1, p_2, \ldots, p_n$.

LEMMA 13.1. *Let $C \subset \mathbb{P}^r$ be a hyperelliptic curve of degree d and genus g; write S for the surface obtained by taking the union of all lines joining pairs of points on C conjugate under the hyperelliptic involution. Then S is a surface of degree*

$$\deg S = d - g - 1.$$

PROOF. Let $\Lambda \subset \mathbb{P}^r$ be a general subspace of codimension 2. Write $\pi\colon C \to \mathbb{P}^1$ for the map induced by projection from Λ; by construction, this is a map of degree d. Similarly, write $\theta\colon C \to \mathbb{P}^1$ for the hyperelliptic map.

Then the points of intersection of Λ with S correspond to pairs of distinct points $(x, y) \in C \times C$, with $\theta(x) = \theta(y)$ and $\pi(x) = \pi(y)$. Equivalently, they correspond to the nodes of the image of C under the map $(\theta, \pi)\colon C \to \mathbb{P}^1 \times \mathbb{P}^1$. This image curve is, by construction, of bidegree $(d, 2)$; in particular, its arithmetic genus is $(d-1)(2-1) = d-1$. The number of nodes is therefore $(d-1)-g = d-g-1 = \deg S$, as desired. $\square$

COROLLARY 13.2. *If $C \subset \mathbb{P}^r$ is of genus 2 and degree $r+2$, then the above surface S has degree*

$$\deg S = r - 1.$$

PROOF. Note that every curve of genus 2 is hyperelliptic, so we may apply apply the previous lemma. $\square$

LEMMA 13.3. *We have*

$$\chi(N_C) = r^2 + 2r + 5.$$

PROOF. We simply calculate

$$\begin{aligned}\chi(N_C) &= (r+1)d - (r-3)(g-1)\\ &= (r+1)(r+2) - (r-3)(2-1)\\ &= r^2 + 2r + 5.\end{aligned}$$

$\square$

To show that interpolation does not hold in the cases of Eq. (13.1), we study the short exact sequence

$$0 \to N_{C/S} \to N_C \to N_S|_C \to 0.$$

By Theorem 4.14 part a, for N_C to satisfy interpolation, it is necessary for

$$\chi(N_{C/S}) \le \frac{\chi(N_C) + r - 2}{r - 1} = \frac{r^2 + 3r + 3}{r - 1}.$$

For $r \in \{3, 4, 5\}$, the right-hand side is strictly less than 11. It is therefore sufficient to observe:

PROPOSITION 13.4. *We have*

$$\chi(N_{C/S}) = 11.$$

So in particular, the bundles N_C do not satisfy interpolation for

$$(d, g, r) \in \{(5, 2, 3), (6, 2, 4), (7, 2, 5)\}.$$

PROOF. Our first task is to understand the intersection theory on S. We have two natural divisor classes F and H on S, where H is the class of a hyperplane section and F is the class of a line connecting two points on C which are sent to each other under the hyperelliptic involution. Then S is a $\mathbb{P}^1$-bundle over $\mathbb{P}^1$, with F the class of a fiber. As $H \cdot F = 1$, this shows that H and F generate the Picard group of S. We know that

$$F \cdot F = 0, \quad F \cdot H = 1, \quad \text{and} \quad H \cdot H = r - 1$$

(as S is a surface of degree $r - 1$).

Now assume C has the class $a \cdot F + b \cdot H$. We know that $C \cdot F = 2$ and $C \cdot H = r+2$. This gives us the equations $b = 2$ and $a + (r-1)b = r+2$, so $(a, b) = (4-r, 2)$. Now this implies that $N_{C/S} = \mathcal{O}(C)|_C$ has degree given by $C \cdot C = 4(4-r) + 4(r-1) = 12$, so by Riemann-Roch $\chi(N_{C/S}) = 11$. □

PROPOSITION 13.5. *Assume $r \in \{3, 4, 5\}$. There exists a non-degenerate curve of genus 2 and degree $r + 2$ through n general points if and only if there exists a ruled non-degenerate surface of degree $r - 1$ through n general points.*

PROOF. We first show, using a dimension-counting argument, that every non-degenerate ruled surface of degree $r-1$ in $\mathbb{P}^r$ contains a curve of genus 2 and degree $r + 2$.

By a result of [**5**], any ruled non-degenerate surface of degree $r - 1$ in $\mathbb{P}^r$ must be a rational normal scroll. By Lemma 2.6 of [**3**], the space of such surfaces has dimension

$$(r + 1)r + r - 6 = r^2 + 2r - 6.$$

Now note that we have a rational map from the space of non-degenerate curves of genus 2 and degree $r+2$ to this space of surfaces, given by our earlier construction of a ruled surface associated to a hyperelliptic curve embedded into projective space.

We previously calculated that $\chi(N_{C/S}) = 11$. Furthermore, as the degree of $N_{C/S}$ was 12, we must have $H^1(N_{C/S}) = 0$. As the space of possible C is irreducible, this implies that the dimension of a generic fiber of this rational map is 11. But we calculated the dimension of the space of possible C to be

$$r^2 + 2r + 5 = r^2 + 2r - 6 + 11,$$

so this map must be dominant.

Now assume that we have a general S through n general points. We have just shown that the space of possible C for a general S is 11-dimensional, and as C is a divisor in S, this space must be the projectivized space of global sections of an appropriate line bundle. Thus, if $n \leq 11$, then there is a curve passing through our n points on the surface. So take $n > 11$. Then

$$r^2 + 2r - 6 \geq n(r-2) \geq 11(r-2) \quad \Rightarrow \quad r^2 - 9r + 16 \geq 0,$$

which is false for $r = 3, 4, 5$. So in this case there is neither an S nor a C through n general points. $\square$

Proof of Theorem 1.3. Except in the cases $(d, g, r) \in \{(5,2,3), (6,2,4), (7,2,5)\}$, this is immediate from Theorem 1.2. It thus remains to consider the cases $(d, g, r) \in \{(5,2,3), (6,2,4), (7,2,5)\}$, in which case we want to prove a non-special curve C of degree d and genus g in $\mathbb{P}^r$ passes through n general points if and only if $n \leq 9$. In these cases, we appeal to Theorem 13.5, which reduces our problem to showing that a ruled surface S of degree $r - 1$ passes through n general points if and only if $n \leq 9$.

For $r = 3$, such a surface is a quadric, and it is easy to see that there is a quadric through n general points if and only if $n \leq 9$. For $r > 3$, it is known by [**3**] (last paragraph of Section 5) that there are $(r-2)(r-3) \neq 0$ scrolls through $r + 4$ general points that meet a general $r - 4$ plane, so for $r \in \{4, 5\}$ we also have that there is a scroll through n points if and only if $n \leq 9$. $\square$

APPENDIX A

Remainder of Proof of Theorem 1.2

In this appendix, we will show by a purely combinatorial argument that any case of Theorem 1.2 can be reduced, using our inductive constructions in Chapter 10, to one of the base cases considered in Chapter 11.

1. Compatibility with (9.1)

To avoid duplicating work, we begin in this section by determining when each of our inductive constructions preserves the condition of Eq. (9.1)—i.e. what additional condition (in terms of d, g, r, and n), in addition to Eq. (9.1) for $(d, g, r; n)$, implies Eq. (9.1) for the various $(d', g', r'; n')$ appearing in the results of Chapter 10. We will restrict ourselves only to those lemmas (and cases thereof) which will be used most commonly; the others will be addressed later, as we invoke them.

While in general we will merely substitute $(d', g', r'; n')$ into Eq. (9.1) (a task we leave to the reader), we can obtain a better result in the case of Lemma 10.5:

LEMMA A.1. *In Lemma* 10.5, *the tuple* $(d-1, g, r-1, n')$ *always satisfies Eq.* (9.1).

PROOF. If $\sum_{i,j,k} kn_{ij}^k < r-2$, then we want to verify

$$\sum_{i,j,k}((r-3)j-k)\cdot n_{ij}^k \leq 2(d-1)+2g-(r-1)-2 = 2d+2g-r-3.$$

Similarly, for $\sum_{i,j,k} kn_{ij}^k = r-2$, we want to verify

$$\sum_{i,j,k}((r-3)j)\cdot n_{ij}^k \leq 2(d-1)+2g-(r-1)-2 = 2d+2g-r-3.$$

Since $\sum_{i,j,k}((r-3)j-k)\cdot n_{ij}^k \leq \sum_{i,j,k}((r-3)j)\cdot n_{ij}^k$, it suffices to verify in all cases that

$$\sum_{i,j,k}((r-3)j)\cdot n_{ij}^k \leq 2d+2g-r-3.$$

But we have

$$\begin{aligned}\sum_{i,j,k}((r-3)j)\cdot n_{ij}^k &= \sum_{i,j,k}((r-2)i+(r-3)j-k)\cdot n_{ij}^k - (r-2)\cdot\sum_{i,j,k} in_{ij}^k + \sum_{i,j,k} kn_{ij}^k \\ &\leq 2d+2g-r-2-(r-2)\cdot\sum_{i,j,k} in_{ij}^k + r-2 \\ &= 2d+2g-4-(r-2)\cdot\sum_{i,j,k} in_{ij}^k.\end{aligned}$$

We are thus done if $\sum_{i,j,k} in_{ij}^k \geq 2$. Since $n_{ij}^k = 0$ unless $i \geq j$, we can thus assume $\sum_{i,j,k} jn_{ij}^k \leq 1$. In this case,

$$\sum_{i,j,k}((r-3)j) \cdot n_{ij}^k \leq r-3 \leq r-3+2(d+g-r) = 2d+2g-r-3. \qquad \square$$

For the other commonly used lemmas in Chapter 10, we simply substitute (d', g', r', n') into Eq. (9.1), and rearrange. Collecting these results (together with the result of Theorem A.1), we obtain the following table of extra conditions necessary for the (d', g', r', n') to satisfy Eq. (9.1).

Result	$\boldsymbol{\sum_{i,j,k} kn_{ij}^k}$	**Condition**
Theorem 10.2	$\leq r-3$	$\sum_{i,j,k}((r-2)i+(r-3)j-k)n_{ij}^k \leq 2d+2g-3r$
Theorem 10.3	$\geq r-3$	$\sum_{i,j,k}((r-2)i+(r-3)j-k)n_{ij}^k \leq 2d+2g-3r$
Theorem 10.5	Arbitrary	No Condition
Theorem 10.6	$\leq r-4$	$\sum_{i,j,k}((r-3)j-k)n_{ij}^k \leq 2d+2g-3r$
Theorem 10.6	$\geq r-3$	$\sum_{i,j,k}((r-3)j-k)n_{ij}^k \leq 2d+2g-4r+2$
Theorem 10.7	$\leq r-3$	$\sum_{i,j,k}((r-2)i+(r-3)j-k)n_{ij}^k \leq 2d+2g-r-4$
Theorem 10.7	$=r-2$	$\sum_{i,j,k}((r-2)i+(r-3)j-k)n_{ij}^k \leq 2d+2g-r-4$ and $\sum_{i,j,k}((r-3)i+(r-4)j)n_{ij}^k \leq 2d+2g-r-3$
Theorem 10.10	Arbitrary	$\sum_{i,j,k}((r-2)i+(r-3)j)n_{ij}^k \leq 2d+2g-r-2$

We now further consider the special case where

$$\sum_{i,j,k}((r-2)i+(r-3)j-k) \cdot n_{ij}^k \leq 2d+2g-3r+2.$$

In this case, we consider Lemmas 10.6, 10.7, and 10.10. For these lemmas, we use this relation to simplify our previous inequalities; these alternate inequalities are collected below in the following table:

Result	$\boldsymbol{\sum_{i,j,k} kn_{ij}^k}$	**Condition**
Theorem 10.6	$\leq r-4$	$\sum_{i,j,k} in_{ij}^k \geq 1$
Theorem 10.6	$\geq r-3$	$\sum_{i,j,k} in_{ij}^k \geq 2$
Theorem 10.7	Arbitrary	No Condition
Theorem 10.10	Arbitrary	No Condition

2. Interpolation for rational curves

In this section, we prove that for rational curves, (9.1) is in fact a sufficient condition for N'_C to satisfy interpolation. In the case of no marked points ($N'_C = N_C$), this result was obtained independently by both Sacchiero [**10**] and Ran [**9**]; however, our proof here will be independent. We will do this by induction on the degree of C.

LEMMA A.2. *Assume all tuples $(d', 0, r'; n')$ satisfying (9.1) and $d' < d$ are good. If*

$$\sum_{i,j,k} ((r-2)i + (r-3)j - k) \cdot n_{ij}^k \leq 2d - 3r + 1,$$

then $(d, 0, r; n)$ is good.

PROOF. First we note that the given inequality implies

$$-(r-2) \leq \sum_{i,j,k} ((r-2)i + (r-3)j - k) \cdot n_{ij}^k \leq 2d - 3r + 1 \implies d \geq \frac{2r+1}{2} > r = g + r.$$

Next, for any $(i, j; k)$ with $i \geq 1$ and $j \geq 0$, we have

$$r - 1 - i - 2j - k \leq (r-2)i + (r-3)j - k;$$

which in turn implies

$$\sum_{\substack{i,j,k \\ (i,j) \neq (0,0)}} (r - 1 - i - 2j - k) n_{ij}^k \leq \sum_{\substack{i,j,k \\ (i,j) \neq (0,0)}} ((r-2)i + (r-3)j - k) \cdot n_{ij}^k.$$

Additionally, since $\sum_{i,j,k} k n_{ij}^k \leq r - 2$, we have

$$\sum_k (r - 1 - k) n_{00}^k \leq \sum_k ((r-1)k - k) n_{00}^k \leq (r-1)(r-2) - \sum_k k n_{00}^k.$$

Adding these together, we obtain

$$\begin{aligned}
\sum_{i,j,k} (r - 1 - i - 2j - k) n_{ij}^k &\leq (r-1)(r-2) + \sum_{i,j,k} ((r-2)i + (r-3)j - k) \cdot n_{ij}^k \\
&\leq (r-1)(r-2) + 2d - 3r + 1 \\
&= (r-1)(r+1) + 2d - 2 - (6r - 6) \\
&\leq (r-1)d + 2d - 2 \\
&= (r+1)d - 2.
\end{aligned}$$

Interpolation thus follows from Lemma 10.7 (c.f. Appendix 1). □

THEOREM A.3. *All tuples $(d, 0, r; n)$ satisfying (9.1) are good.*

PROOF. Assume otherwise. Take a counterexample with minimal d. If

$$\sum_{i,j,k} ((r-2)i + (r-3)j - k) \cdot n_{ij}^k \leq 2d - 3r + 1$$

then A.2 gives a contradiction. But if

$$\sum_{i,j,k} ((r-2)i + (r-3)j - k) \cdot n_{ij}^k \geq 2d - 3r + 2$$

then Lemma 10.5 gives a contradiction. □

3. Space curves

In this section, we prove that following result.

THEOREM A.4. *The tuple $(d, g, 3; n)$ is good provided it satisfies (9.1), and does not lie in one of two infinite families:*

- $\sum_{i,j,k} jn_{ij}^k = \sum kn_{ij}^k = 0$ *with* $g \neq 0$, *and*
$$\sum in_{ij} = 2d + 2g - 14.$$
- $\sum_{i,j,k} kn_{ij}^k = 1$ *with* $g \neq 0$, *and*
$$\sum in_{ij} = 2d + 2g - 9.$$

As Theorem A.3 takes care of the $g = 0$ case, we will assume that $g \neq 0$ for the rest of this section. Also, we note that (9.1) can be rewritten for $r = 3$ as

$$\sum_{i,j,k} (i - k) \cdot n_{ij}^k \leq 2d + 2g - 5. \tag{A.1}$$

We prove this theorem by induction on d; and for fixed values of d by induction on the number of marked points.

LEMMA A.5. *Suppose that Theorem A.4 holds for $(d', g', 3; n')$ for all $d' < d$. Then Theorem A.4 holds for $(d, g, 3; n)$ provided that*

$$g > 0, \quad \sum_{i,j,k} kn_{ij}^k = 1, \quad \textit{and} \quad \sum_{i,j,k} in_{ij}^k \leq 2d + 2g - 9.$$

PROOF. By the assumption of Theorem A.4, we have $\sum_{i,j,k} in_{ij}^k \neq 2d+2g-9$; in particular, our assumption in fact implies

$$\sum_{i,j,k} in_{ij}^k \leq 2d + 2g - 10.$$

Applying Lemma 10.2, it suffices to show $(d - 1, g - 1, 3; n')$ is good, where

$$(n')_{ij}^k = \begin{cases} \sum_\ell n_{ij}^\ell & \text{if } k = 0 \text{ and } (i, j; k) \notin \{(0, 0; 0), (1, 1; 0)\}; \\ 1 + \sum_\ell n_{ij}^\ell & \text{if } (i, j; k) = (1, 1; 0); \\ 0 & \text{else.} \end{cases}$$

By our inductive hypothesis, it is sufficient to see that $(d-1, g-1, r; n')$ satisfies (9.1) and does not lie in either of the above infinite families. For (9.1), we want

$$1 + \sum in_{ij}^k \leq 2(d - 1) + 2(g - 1) - 5 = 2d + 2g - 9,$$

which is precisely what we observed above. To see it does not lie in either of our infinite families, we note that $(d - 1, g - 1, r; n')$ satisfies $\sum_{i,j,k} k(n')_{ij}^k = 0$, but $\sum_{i,j,k} j(n')_{ij}^k \neq 0$. □

LEMMA A.6. *Suppose that Theorem A.4 holds for $(d', g', 3; n')$ for all $d' < d$; and also when $d = d'$ and n' has fewer marked points than n. Then Theorem A.4 holds for $(d, g, 3; n)$ provided that*

$$g > 0, \quad \sum_{i,j,k} kn_{ij}^k = 0, \quad \textit{and} \quad \sum_{i,j,k} in_{ij}^k \leq 2d + 2g - 9.$$

PROOF. Again, Lemma 10.2 implies that it suffices to show $(d-1, g-1, 3; n')$ is good, where

$$(n')^k_{ij} = \begin{cases} n^k_{ij} & \text{if } (i,j;k) \neq (1,1;1); \\ n^1_{11} + 1 & \text{if } (i,j;k) = (1,1;1). \end{cases}$$

If $\sum_{i,j,k} in^k_{ij} \neq 2d + 2g - 14$, then we have

$$\sum_{i,j,k} i(n')^k_{ij} = 1 + \sum_{i,j,k} in^k_{ij} \neq 1 + 2d + 2g - 14 = 2(d-1) + 2(g-1) - 9;$$

and so because $\sum_{i,j,k} k(n')^k_{ij} = 1$, our inductive hypothesis implies $(d-1, g-1, 3; n')$ is good, subject to the inequality

$$\sum_{i,j,k} in^k_{ij} \leq 2(d-1) + 2(g-1) - 5 = 2d + 2g - 9.$$

It thus remains to consider the case when $\sum_{i,j,k} in^k_{ij} = 2d + 2g - 14$. But in this case, our assumptions in Theorem A.4 imply $\sum_{i,j,k} jn^k_{ij} > 0$. Moreover, by assumption, $\sum_{i,j,k} kn^k_{ij} = 0$. Consequently, we must have some point of type $(\ell, m; 0)$ with $m \neq 0$. Applying Lemma 10.9, we conclude that $(d, g, 3; n)$ is good provided that $(d, g, 3; n'')$ is good, where

$$(n'')^k_{ij} = \begin{cases} n^k_{ij} - 1 & \text{if } (i,j;k) = (\ell, m; 0); \\ n^k_{ij} & \text{else.} \end{cases}$$

Since $\sum_{i,j,k} k(n'')^k_{ij} = \sum_{i,j,k} kn^k_{ij} = 0$, it is sufficient, by our inductive assumption, to note that

$$\sum_{i,j,k} i(n'')^k_{ij} = \sum_{i,j,k} in^k_{ij} - \ell = 2d + 2g - 14 - \ell < 2d + 2g - 14.$$

(Above, we used that $\ell \geq m > 0$, so $\ell \neq 0$.) □

LEMMA A.7. *Suppose that Theorem* A.4 *holds for* $(d', g', 3; n')$ *for all* $d' < d$*; and also when* $d = d'$ *and* n' *has fewer marked points than* n*. Then Theorem* A.4 *holds for* $(d, g, 3; n)$ *provided that*

$$\sum_{i,j,k} (i - k) \cdot n^k_{ij} \leq 2d + 2g - 10.$$

PROOF. Since $\sum_{i,j,k} kn^k_{ij} \leq 1$, our given inequality implies

$$\sum_{i,j,k} in^k_{ij} \leq 2d + 2g - 9.$$

If $g = 0$, then the result follows from Theorem A.3. Otherwise, if $g > 0$, then the result follows from Lemma A.5 or A.6, according to whether $\sum_{i,j,k} kn^k_{ij} = 0$ or $\sum_{i,j,k} kn^k_{ij} = 1$. □

Theorem A.4 then follows from combining Lemmas A.7, 11.2, and 10.5.

COROLLARY A.8. *Theorem* 1.2 *holds for* $r = 3$.

PROOF. If $n_{ij}^k = 0$ for all $(i, j; k)$, then Theorem A.4 implies N_C satisfies interpolation, unless $g \neq 0$ and

$$0 = 2d + 2g - 14 \quad \Rightarrow \quad d + g = 7.$$

Since $d \geq g + 3$, this means either $(d, g) = (5, 2)$ or $(d, g) = (6, 1)$. The case of $(d, g) = (5, 2)$ is excluded by the assumption of Theorem 1.2; it thus suffices to show N_C satisfies interpolation for $(d, g, r) = (6, 1, 3)$.

In this case, we apply Lemma 10.7, which implies the desired result so long as $(5, 1, 3; \mathbf{0})$ and $(5, 1, 2; \mathbf{0})$ are both good. But these follow from Theorem A.4 and Lemma 11.1 respectively. □

4. Curves in low dimensional projective spaces

In this section, we study curves in $\mathbb{P}^r$, where $4 \leq r \leq 11$. Combined with the results of the previous section for curves in $\mathbb{P}^3$, this establishes Theorem 1.2 for $r \leq 11$. Note that this range includes all the counterexamples to interpolation listed in Theorem 1.2 — as well as the counterexample-free dimension $r = 11$, which will serve (along with Theorem A.3) as the base case of our inductive argument for higher-dimensional projective spaces.

DEFINITION A.9. We say that (d, g, r) is *excellent* if $(d, g, r; n)$ is good for every n satisfying Eq. (9.1).

In these terms, our basic goal is to demonstrate the following.

THEOREM A.10. *Let $r \geq 4$, and suppose that $d + g \geq 2r - 1$ and*

$$(d - 1, g - 1, r), \quad (d - 1, g, r - 1), \quad \textit{and} \quad (d - 2, g - 1, r - 1)$$

are all excellent. Then (d, g, r) is excellent.

PROOF. If $g = 0$, then the result follows from Theorem A.3; we thus suppose $g > 0$. If

$$\sum_{i,j,k} ((r - 2)i + (r - 3)j - k) \cdot n_{ij}^k \leq 2d + 2g - 3r,$$

then the desired result follows from Lemma 10.2 if $\sum_{i,j,k} kn_{ij}^k < r - 2$, and from Lemma 10.3 if $\sum_{i,j,k} kn_{ij}^k = r - 2$. On the other hand, if

$$\sum_{i,j,k} ((r - 2)i + (r - 3)j - k) \cdot n_{ij}^k \geq 2d + 2g - 3r + 2,$$

then the desired result follows from Lemma 10.5. It thus remains to consider the case where

$$\sum_{i,j,k} ((r - 2)i + (r - 3)j - k) \cdot n_{ij}^k = 2d + 2g - 3r + 1.$$

If $\sum_{i,j,k} in_{ij}^k \geq 2$, then the desired result follows Lemma 10.6. We are left with the case $\sum_{i,j,k} in_{ij}^k \leq 1$. If $\sum_{i,j,k} in_{ij}^k = 1$, we have $\sum_{i,j,k} jn_{ij}^k \leq 1$; and we may assume $\sum_{i,j,k} kn_{ij}^k \geq r - 3$ (since otherwise we may again apply Lemma 10.6). Consequently,

$$2d + 2g - 3r + 1 = \sum_{i,j,k} ((r - 2)i + (r - 3)j - k) \cdot n_{ij}^k \leq (r - 2) + (r - 3) - (r - 3) = r - 2.$$

Similarly, if $\sum_{i,j,k} in_{ij}^k = 0$, then we have $\sum_{i,j,k} jn_{ij}^k = 0$ as well, which gives

$$2d + 2g - 3r + 1 = \sum_{i,j,k} ((r-2)i + (r-3)j - k) \cdot n_{ij}^k \leq 0 + 0 - 0 = 0.$$

Either way,

$$2d + 2g - 3r + 1 \leq r - 2.$$

But this contradicts our assumption that $d + g \geq 2r - 1$. □

PROPOSITION A.11. *All tuples $(d, g, 4)$ with $d \geq g + 4$ with $d + g \geq 11$ are excellent. In addition, Theorem 1.2 holds for $r = 4$.*

PROOF. We argue by induction on $d + g$. It is a finite computation to verify the proposition in the range $d + g \leq 16$ (see Appendix B). For the inductive step, we thus suppose $d + g \geq 17$.

In particular, unless $g = 0$ (in which case the result follows from Theorem A.3), $(d - 1, g - 1, 4)$ is excellent by our inductive hypothesis. As in Theorem A.10, this implies the desired result when

$$\sum_{i,j,k} ((r-2)i + (r-3)j - k) \cdot n_{ij}^k \leq 2d + 2g - 3r.$$

We next consider the case when

$$\sum_{i,j,k} ((r-2)i + (r-3)j - k) \cdot n_{ij}^k \geq 2d + 2g - 3r + 2.$$

In this case, Theorem 10.5 implies the desired result provided that $(d - 1, g, 3; n')$ is good, where if

$$\sum_{i,j,k} kn_{ij}^k < r - 2,$$

then

$$(n')_{ij}^k = \begin{cases} \sum_\ell n_{\ell i}^k & \text{if } j = 0 \text{ and } (i, j; k) \neq (0, 0; 0), \\ 0 & \text{else;} \end{cases}$$

and if

$$\sum_{i,j,k} kn_{ij}^k = r - 2,$$

then

$$(n')_{ij}^k = \begin{cases} \sum_{\ell,m} n_{\ell i}^m & \text{if } j = k = 0 \text{ but } i \neq 0, \\ 0 & \text{else.} \end{cases}$$

In either case, we know (c.f. Appendix 1) that Eq. (9.1) is satisfied; it thus remains to check that neither case falls into the exceptional families of Theorem A.4. But in either case, we have

$$\sum_{i,j,k} i(n')_{ij}^k = \sum_{i,j,k} jn_{ij}^k;$$

so it remains to show

$$\sum_{i,j,k} jn_{ij}^k \leq 2(d-1) + 2g - 15 = 2d + 2g - 17. \tag{A.2}$$

We will return to this after first considering the case where

$$\sum_{i,j,k}((r-2)i+(r-3)j-k)\cdot n_{ij}^k = 2d+2g-3r+1.$$

As in Theorem A.10, our assumption that $d+g \geq 2r-1 = 7$ implies that either $\sum_{i,j,k} in_{ij}^k \geq 2$, or $\sum_{i,j,k} in_{ij}^k = 1$ and $\sum_{i,j,k} kn_{ij}^k \leq r-4$; either way, Lemma 10.6 implies the desired result provided that $(d-2, g-1, r-1; n')$ is good, where if

$$\sum_{i,j,k} kn_{ij}^k < r-3,$$

then

$$(n')_{ij}^k = \begin{cases} \sum_\ell n_{\ell i}^k & \text{if } j=0 \text{ and } (i,j;k) \notin \{(0,0;0),(2,0,1)\}, \\ 1+\sum_\ell n_{\ell i}^k & \text{if } (i,j;k) = (2,0,1), \\ 0 & \text{else;} \end{cases}$$

and if

$$\sum_{i,j,k} kn_{ij}^k = r-3,$$

then

$$(n')_{ij}^k = \begin{cases} \sum_{\ell,m} n_{\ell i}^m & \text{if } j=k=0 \text{ and } i \notin \{0,2\}, \\ 1+\sum_{\ell,m} n_{\ell i}^m & \text{if } j=k=0 \text{ and } i=2, \\ 0 & \text{else;} \end{cases}$$

and finally if

$$\sum_{i,j,k} kn_{ij}^k = r-2,$$

then

$$(n')_{ij}^k = \begin{cases} \sum_{\ell,m} n_{\ell i}^m & \text{if } j=k=0 \text{ and } i \neq 0, \\ 1 & \text{if } (i,j;k) = (2,0,1), \\ 0 & \text{else.} \end{cases}$$

In either case, we know (c.f. Appendix 1) that Eq. (9.1) is satisfied; it thus remains to check that neither case falls into the exceptional families of Theorem A.4. But in either case, we have

$$\sum_{i,j,k} i(n')_{ij}^k = 2+\sum_{i,j,k} jn_{ij}^k;$$

so it remains to show

$$2+\sum_{i,j,k} jn_{ij}^k \leq 2(d-2)+2(g-1)-15 = 2d+2g-21,$$

or equivalently, that

$$\sum_{i,j,k} jn_{ij}^k \leq 2d+2g-23. \tag{A.3}$$

Since Eq. (A.3) visibly implies Eq. (A.2), we conclude that to verify this proposition, it suffices to prove Eq. (A.3). For this, we calculate

$$\begin{aligned}\sum_{i,j,k}((r-2)i+(r-3)j-k)\cdot n_{ij}^k &= \sum_{i,j,k}(2i+j-k)\cdot n_{ij}^k\\ &\geq \sum_{i,j,k}(3j-k)\cdot n_{ij}^k\\ &\geq 3\cdot\sum_{i,j,k} jn_{ij}^k-(r-2).\end{aligned}$$

Using Eq. (9.1), this implies

$$3\cdot\sum_{i,j,k} jn_{ij}^k - r + 2 \leq 2d+2g-r-2;$$

or upon rearrangement,

$$\sum_{i,j,k} jn_{ij}^k \leq \frac{2d+2g-4}{3}.$$

It thus suffices to note that

$$\frac{2d+2g-4}{3} \leq 2d+2g-23;$$

which follows immediately from our assumption that $d+g \geq 17$. □

PROPOSITION A.12. *All tuples $(d, g, 5)$ with $d \geq g+5$ and $d+g \geq 14$ are excellent. In addition, Theorem* 1.2 *holds for $r = 5$.*

PROOF. We argue by induction on $d+g$. It is a finite computation to verify the proposition in the range $d+g \leq 15$ (see Appendix B). For the inductive step, we thus suppose $d+g \geq 16$.

In particular, unless $g = 0$ (in which case the result follows from Theorem A.3), $(d-1, g-1, 5)$ is excellent by our inductive hypothesis. Moreover, by Proposition A.11, both $(d-1, g, 4)$ and $(d-2, g-1, 4)$ are excellent. Theorem A.10 thus implies the desired result. □

PROPOSITION A.13. *All tuples $(d, g, 6)$ with $d \geq g+6$ and $d+g \geq 13$ are excellent. In addition, Theorem* 1.2 *holds for $r = 6$.*

PROOF. Again, we argue by induction on $d+g$. It is a finite computation to verify the proposition in the range $d+g \leq 16$ (see Appendix B). For the inductive step, we thus suppose $d+g \geq 17$.

In particular, unless $g = 0$ (in which case the result follows from Theorem A.3), $(d-1, g-1, 6)$ is excellent by our inductive hypothesis. Moreover, by Proposition A.12, both $(d-1, g, 5)$ and $(d-2, g-1, 5)$ are excellent. Theorem A.10 thus implies the desired result. □

PROPOSITION A.14. *All tuples $(d, g, 7)$ with $d \geq g+7$ and $d+g \geq 14$ are excellent. In addition, Theorem* 1.2 *holds for $r = 7$.*

PROOF. Again, we argue by induction on $d+g$. It is a finite computation to verify the proposition in the range $d+g \leq 15$ (see Appendix B). For the inductive step, we thus suppose $d+g \geq 16$.

In particular, unless $g = 0$ (in which case the result follows from Theorem A.3), $(d-1, g-1, 7)$ is excellent by our inductive hypothesis. Moreover, by Proposition A.13, both $(d-1, g, 6)$ and $(d-2, g-1, 6)$ are excellent. Theorem A.10 thus implies the desired result. □

PROPOSITION A.15. *All tuples $(d, g, 8)$ with $d \geq g+8$ are excellent. (In particular, Theorem 1.2 holds for $r = 8$.)*

PROOF. Again, we argue by induction on $d+g$. It is a finite computation to verify the proposition in the range $d+g \leq 16$ (see Appendix B). For the inductive step, we thus suppose $d+g \geq 17$.

In particular, unless $g = 0$ (in which case the result follows from Theorem A.3), $(d-1, g-1, 8)$ is excellent by our inductive hypothesis. Moreover, by Proposition A.13, both $(d-1, g, 7)$ and $(d-2, g-1, 7)$ are excellent. Theorem A.10 thus implies the desired result. □

PROPOSITION A.16. *All tuples (d, g, r) with $d \geq g+r$ and $9 \leq r \leq 11$ are excellent. (In particular, Theorem 1.2 holds for $9 \leq r \leq 11$.)*

PROOF. Again, we argue by induction on $d+g$. By Theorem A.10, it is sufficient to check the range $d+g \leq 2r-2$. But this a finite computation (c.f. Appendix B). □

5. Curves in high dimensional projective spaces

In this section we study curves in $\mathbb{P}^r$, where $r \geq 12$. In order to state our main result, we will need the following definition:

DEFINITION A.17. Suppose that

$$\sum_{i,j,k} (i+j) n_{ij}^k \leq 3.$$

Then we define $\delta(n)$ according to the following table.

$\sum_{i,j,k} i n_{ij}^k$	$\sum_{i,j,k} j n_{ij}^k$	$\delta(n)$
0	0	2
1	0	3
1	1	5
2	0	4
2	1	5
3	0	4

Our main result will be the following theorem, which we will prove by induction on r.

THEOREM A.18. *The tuple $(d, g, r; n)$ is good if $r \geq 11$ and $d \geq g+r$, unless either*

$$\sum_{i,j,k} i n_{ij}^k = \sum_{i,j,k} j n_{ij}^k = 1, \quad \sum_{i,j,k} k n_{ij}^k = r-2, \quad \text{and} \quad d+g = 2r-2;$$

or

$$\sum_{i,j,k} (i+j) \cdot n_{ij}^k \leq 3, \qquad \sum_{i,j,k} k n_{ij}^k = 4r - 2d - 2g - \delta(n) > \frac{r}{2}, \quad \text{and}$$
$$d+g+r \equiv \delta(n)+2 \text{ or } \delta(n)+4 \mod 5.$$

Note that Theorem A.16 implies the Theorem A.18 for $r = 11$; this will serve as the base case of our induction. For our inductive step, we will therefore suppose $r \geq 12$.

Before proving Theorem A.18, we first deduce two useful corollaries. These corollaries assert that certain tuples $(d, g, r; n)$ are good, and only require the truth of Theorem A.18 for tuples $(d, g, r; n')$ which satisfy $\sum_{i,j,k} (n')^k_{ij} \leq \sum_{i,j,k} n^k_{ij}$. These corollaries can therefore be used in our inductive argument. We begin with the following lemma.

LEMMA A.19. *The inequalities of Theorem 10.10 and Theorem 10.7 are satisfied provided that*

$$\sum_{i,j,k} (i+k) \cdot n^k_{ij} \leq \frac{3r^2 - 3r - 4}{2r - 4} - \frac{r-5}{2r-4}(d+g).$$

PROOF. Subject to the given inequality,

$$\begin{aligned}
\sum_{i,j,k} (r - 1 - i - 2j - k) \cdot n^k_{ij} &\leq (r-2) \cdot \sum_{i,j,k} (i+k) \cdot n^k_{ij} \\
&\leq (r-2) \cdot \left(\frac{3r^2 - 3r - 4}{2r - 4} - \frac{r-5}{2r-4}(d+g) \right) \\
&= \frac{3r-3}{2} \cdot r - \frac{r-5}{2}(d+g) - 2 \\
&\leq \frac{3r-3}{2}(d-g) - \frac{r-5}{2}(d+g) - 2 \\
&= (r+1)d - (2r-4)g - 2. \qquad \square
\end{aligned}$$

The following corollary gives a slight strengthening of Theorem A.18, which will be useful for induction: Once we prove it, we may assume the stronger statement given below as our inductive hypothesis, but need only show the weaker statement of Theorem A.18.

COROLLARY A.20. *The tuple $(d, g, r; n)$ is good if $r \geq 11$ and $d \geq g + r$, unless either*

$$\sum_{i,j,k} i n^k_{ij} = \sum_{i,j,k} j n^k_{ij} = 1, \quad \sum_{i,j,k} k n^k_{ij} = r - 2, \quad \text{and} \quad d + g = 2r - 2;$$

or

$$\sum_{i,j,k} (i+j) \cdot n^k_{ij} \leq 3, \qquad \sum_{i,j,k} k n^k_{ij} = 4r - 2d - 2g - \delta(n) > \frac{r+3}{2}, \quad \text{and}$$
$$d + g + r \equiv \delta(n) + 2 \text{ or } \delta(n) + 4 \mod 5.$$

PROOF. For $r = 11$, this follows from Theorem A.16; we thus assume $r \geq 12$. Applying Theorem A.18, it suffices to consider the case where

$$\text{(A.4)} \quad \sum_{i,j,k} (i+j) \cdot n^k_{ij} \leq 3 \quad \text{and} \quad \frac{r+1}{2} \leq \sum_{i,j,k} k n^k_{ij} = 4r - 2d - 2g - \delta(n) \leq \frac{r+3}{2}.$$

By induction, it is sufficient to show that in such a case, we can always apply Theorem 10.10. For this, we first need to know that $n^k_{00} \neq 0$ for some k. But

$$\sum_{i,j,k} k n^k_{ij} \geq \frac{r+1}{2} \geq \frac{13}{2} > 2 \cdot 3 \geq \sum_{i,j,k} 2(i+j) n^k_{ij} \geq \sum_{\substack{i,j,k \\ (i,j) \neq (0,0)}} k n^k_{ij}.$$

Next we need to check the inequalities of Theorem 10.10. By Theorem A.19 and the results of Appendix 1, this boils down to showing the two inequalities:

$$\sum_{i,j,k} ((r-2)i + (r-3)j - k) \cdot n_{ij}^k \leq 2d + 2g - r - 2,$$

$$\sum_{i,j,k} (i+k) \cdot n_{ij}^k \leq \frac{3r^2 - 3r - 4}{2r-4} - \frac{r-5}{2r-4}(d+g).$$

By our assumption that $\sum_{i,j,k}(i+j)n_{ij}^k \leq 3$, these reduce to:

$$3(r-2) - \sum_{i,j,k} jn_{ij}^k - \sum_{i,j,k} kn_{ij}^k \leq 2d + 2g - r - 2,$$

$$3 + \sum_{i,j,k} kn_{ij}^k \leq \frac{3r^2 - 3r - 4}{2r-4} - \frac{r-5}{2r-4}(d+g).$$

Solving for $d+g$ in Eq. (A.4), we obtain:

$$d + g = \frac{4r - \delta(n) - \sum_{i,j,k} kn_{ij}^k}{2}.$$

Substituting this into the above, it remains to show:

$$3(r-2) - \sum_{i,j,k} jn_{ij}^k - \sum_{i,j,k} kn_{ij}^k \leq 2 \cdot \frac{4r - \delta(n) - \sum_{i,j,k} kn_{ij}^k}{2} - r - 2,$$

$$3 + \sum_{i,j,k} kn_{ij}^k \leq \frac{3r^2 - 3r - 4}{2r-4} - \frac{r-5}{2r-4} \cdot \frac{4r - \delta(n) - \sum_{i,j,k} kn_{ij}^k}{2}.$$

Or upon rearrangement, that

$$\delta(n) \leq 4 + \sum_{i,j,k} jn_{ij}^k,$$

$$(3r-3) \cdot \sum_{i,j,k} kn_{ij}^k \leq 2r^2 + 2r + 16 + (r-5) \cdot \delta(n).$$

The first of these inequalities is clear. For the second, we first note that $\delta(n) \geq 2$; it is thus sufficient to show

$$(3r-3) \cdot \sum_{i,j,k} kn_{ij}^k \leq 2r^2 + 2r + 16 + 2(r-5) = 2r^2 + 4r + 6,$$

or upon rearrangement, that

$$\sum_{i,j,k} kn_{ij}^k \leq \frac{2r^2 + 4r + 6}{3r - 3}.$$

Applying Eq. (A.4), it thus remains to show

$$\frac{r+3}{2} \leq \frac{2r^2 + 4r + 6}{3r - 3},$$

which is clear for $r \geq 12$. □

For convenience, we include the following corollary, giving several more easily-used special cases of Theorem A.20, which will appear in our subsequent inductive argument.

COROLLARY A.21. *The tuple* $(d, g, r; n)$ *is good if* $r \geq 11$ *and* $d \geq g+r$, *provided that* (9.1) *is satisfied and at least one of the following holds:*

(1) *If* $d + g \geq 2r - 1$;
(2) *If* $d + g \geq (7r - 7)/4$ *and we do not have both*

$$\sum_{i,j,k} in_{ij}^k = \sum_{i,j,k} jn_{ij}^k = 1 \quad \textit{and} \quad \sum_{i,j,k} kn_{ij}^k = r - 2;$$

(3) *If* $\sum_{i,j,k} kn_{ij}^k \leq (r + 3)/2$. *In particular, this happens if* $n_{ij}^k = 0$ *for all* $(i, j; k)$.

Condition 3 *in particular implies Theorem* 1.2 *holds for* $r \geq 12$ — *which, combined with the results of Section* 4, *completes the proof of Theorem* 1.2.

PROOF. We begin with Conditions 1 and 2, making use of Theorem A.20: If

$$4r - 2d - 2g - \delta(n) > \frac{r + 3}{2},$$

then in particular we have

$$4r - 2d - 2g - 2 \geq 4r - 2d - 2g - \delta(n) \geq \frac{r + 4}{2};$$

or upon rearrangement,

$$d + g \leq \frac{7r - 8}{4}.$$

We conclude that $(d, g, r; n)$ is good unless

$$d + g \leq \max\left(\frac{7r - 8}{4}, 2r - 2\right) = 2r - 2;$$

and unless $\sum_{i,j,k} in_{ij}^k = \sum_{i,j,k} jn_{ij}^k = 1$ and $\sum_{i,j,k} kn_{ij}^k = r - 2$, that $(d, g, r; n)$ is good unless

$$d + g \leq \frac{7r - 8}{4}.$$

Finally, we consider Condition 3: In this case, it sufficient to note that $r - 2 > (r + 3)/2$. □

PROPOSITION A.22. *If Theorem A.18 holds for* $r' = r - 1$, *then to prove Theorem A.18 in* $\mathbb{P}^r$, *it is sufficient to consider cases where*

$$g > 0 \quad \textit{and} \quad d + g \leq 2r. \tag{A.5}$$

PROOF. The case of $g = 0$ is covered by Theorem A.3, so it suffices to consider the cases where $g > 0$. Next, for $d + g \geq 2r + 1$, we have

$$\begin{aligned} (d - 1) + (g - 1) &\geq 2r - 1, \\ (d - 1) + g &\geq 2(r - 1) - 1, \\ (d - 2) + (g - 1) &\geq 2(r - 1) - 1. \end{aligned}$$

Theorem A.10 therefore implies the desired result by induction on $d + g$. □

For the remainder of this section, we will thus make the assumptions given by Eq. (A.5).

PROPOSITION A.23. *Suppose that Theorem A.18 holds for all $(d', g', r'; n')$ where either $d' < d$, or $d' = d$ and $\sum_{i,j,k}(n')^k_{ij} < \sum_{i,j,k} n^k_{ij}$. Then Theorem A.18 holds for $(d, g, r; n)$ provided that*

$$\sum_{i,j,k}((r-2)i + (r-3)j - k) \cdot n^k_{ij} \leq 2d + 2g - 3r \quad \text{and} \quad \sum_{i,j,k} kn^k_{ij} \geq r - 3.$$

PROOF. The desired result follows from Theorem 10.3: Indeed, the n' appearing in Theorem 10.3 satisfies

$$\sum_{i,j,k} k(n')^k_{ij} \leq 2 \leq \frac{r+3}{2}.$$

□

PROPOSITION A.24. *Suppose that Theorem A.18 holds for all $(d', g', r'; n')$ where either $d' < d$, or $d' = d$ and $\sum_{i,j,k}(n')^k_{ij} < \sum_{i,j,k} n^k_{ij}$. Then Theorem A.18 holds for $(d, g, r; n)$ provided that*

$$\sum_{i,j,k}((r-2)i + (r-3)j - k) \cdot n^k_{ij} \geq 2d + 2g - 3r + 2 \quad \text{and} \quad \sum_{i,j,k} kn^k_{ij} = r - 2.$$

PROOF. The desired result follows from Theorem 10.5: The n' appearing in Theorem 10.5 satisfies

$$\sum_{i,j,k} k(n')^k_{ij} = 0 \leq \frac{r+3}{2}.$$

□

LEMMA A.25. *We have*

$$\sum_{i,j,k}((r-2)i + (r-3)j - k) \cdot n^k_{ij} \geq \frac{2r-5}{2} \cdot \sum_{i,j,k}(i+j)n^k_{ij} - r + 2.$$

PROOF. Because $i \geq j$ whenever $n^k_{ij} \neq 0$, we obtain

$$\begin{aligned}\sum_{i,j,k}((r-2)i + (r-3)j - k) \cdot n^k_{ij} &\geq \frac{(r-2)+(r-3)}{2} \cdot \sum_{i,j,k}(i+j)n^k_{ij} - \sum_{i,j,k} kn^k_{ij} \\ &\geq \frac{2r-5}{2} \cdot \sum_{i,j,k}(i+j)n^k_{ij} - (r-2).\end{aligned}$$

□

LEMMA A.26. *We have*

$$\sum_{i,j,k}(i+j)n^k_{ij} \leq \frac{4(d+g)-8}{2r-5}.$$

PROOF. By Eq. (9.1) together with Theorem A.25,

$$\frac{2r-5}{2} \cdot \sum_{i,j,k}(i+j)n^k_{ij} - r + 2 \leq \sum_{i,j,k}((r-2)i + (r-3)j - k) \cdot n^k_{ij} \leq 2(d+g) - r - 2.$$

Rearranging yields the statement of this lemma. □

LEMMA A.27. *We have*

$$\sum_{i,j,k}((r-2)i + (r-3)j - k) \cdot n^k_{ij} \geq 2d + 2g - 3r + 2$$

provided that

$$\sum_{i,j,k}(i+j) \cdot n^k_{ij} \geq \frac{4(d+g)-4r}{2r-5}.$$

PROOF. We have

$$\sum_{i,j,k}((r-2)i+(r-3)j-k)\cdot n_{ij}^k \geq \frac{2r-5}{2}\cdot\sum_{i,j,k}(i+j)\cdot n_{ij}^k - r + 2;$$

it is therefore sufficient to show

$$\frac{2r-5}{2}\cdot\sum_{i,j,k}(i+j)\cdot n_{ij}^k - r + 2 \geq 2d+2g-3r+2,$$

which is a rearrangement of our assumption. □

LEMMA A.28. *We have*

$$\frac{4(d+g)-4r}{2r-5} \leq 3.$$

PROOF. Upon rearrangement, our desired inequality becomes

$$d+g \leq \frac{10r-15}{4}.$$

Using Theorem A.22, it thus remains to check

$$2r \leq \frac{10r-15}{4},$$

which is immediate for $r \geq 8$. □

PROPOSITION A.29. *Suppose that Theorem A.18 holds for all $(d', g', r'; n')$ where either $d' < d$, or $d' = d$ and $\sum_{i,j,k}(n')_{ij}^k < \sum_{i,j,k} n_{ij}^k$. Then Theorem A.18 holds for $(d, g, r; n)$ if $\sum_{i,j,k}(i+j)\cdot n_{ij}^k \geq 4$.*

PROOF. From Theorem A.27 and Theorem A.28, we obtain

$$\sum_{i,j,k}((r-2)i+(r-3)j-k)\cdot n_{ij}^k \geq 2d+2g-3r+2.$$

If $\sum_{i,j,k} kn_{ij}^k = r-2$, then the result follows from Theorem A.24. Otherwise, to conclude by Theorem 10.5, it suffices to show $(d-1, g, r-1; n')$ satisfies our inductive hypothesis, where

$$(n')_{ij}^k = \begin{cases} \sum_\ell n_{\ell i}^k & \text{if } j = 0 \text{ and } (i,j;k) \neq (0,0;0); \\ 0 & \text{else.} \end{cases}$$

Because $\sum_{i,j,k} j(n')_{ij}^k = 0$, it is sufficient (c.f. Theorem A.21) to check

$$(d-1)+g \geq \frac{7(r-1)-7}{4};$$

or upon rearrangement,

$$d+g \geq \frac{7r-10}{4}.$$

However, by Theorem A.26,

$$4 \leq \sum_{i,j,k}(i+j)\cdot n_{ij}^k \leq \frac{4(d+g)-8}{2r-5};$$

which upon rearrangement yields

$$d+g \geq 2r-3.$$

It is thus sufficient to note that for $r \geq 12$,

$$\frac{7r - 10}{4} \leq 2r - 3. \qquad \square$$

PROPOSITION A.30. *Suppose that Theorem A.18 holds for all $(d', g', r'; n')$ where either $d' < d$, or $d' = d$ and $\sum_{i,j,k} (n')^k_{ij} < \sum_{i,j,k} n^k_{ij}$. Then Theorem A.18 holds for $(d, g, r; n)$ if $\sum_{i,j,k} (i + j) \cdot n^k_{ij} \in \{2, 3\}$ and*

$$\frac{4(d + g) - 4r}{2r - 5} \leq \sum_{i,j,k} (i + j) \cdot n^k_{ij}.$$

In particular, Theorem A.18 holds for $\sum_{i,j,k} (i + j) \cdot n^k_{ij} = 3$.

PROOF. By Theorem A.24, it suffices to consider the case $\sum_{i,j,k} kn^k_{ij} < r - 3$. Moreover, by Theorem A.27, the inequality required for Theorem 10.5 is satisfied. Write

$$\varepsilon = \sum_{i,j,k} jn^k_{ij} \in \{0, 1\}.$$

As in Theorem A.29, it suffices to show $(d - 1, g, r - 1; n')$ satisfies our inductive hypothesis, where

$$(n')^k_{ij} = \begin{cases} \sum_\ell n^k_{\ell i} & \text{if } j = 0 \text{ and } (i, j; k) \neq (0, 0; 0); \\ 0 & \text{else.} \end{cases}$$

Because

$$\sum_{i,j,k} i(n')^k_{ij} = \varepsilon \quad \text{and} \quad \sum_{i,j,k} j(n')^k_{ij} = 0,$$

we have $\delta(n') = 2 + \varepsilon$. Our problem is thus to show that we cannot simultaneously have

$$\sum_{i,j,k} kn^k_{ij} = \sum_{i,j,k} k(n')^k_{ij} = 4(r - 1) - 2(d - 1) - 2g - (2 + \varepsilon) > \frac{(r - 1) + 3}{2},$$
$$(d - 1) + g + (r - 1) \equiv 4 + \varepsilon \text{ or } 1 + \varepsilon \mod 5.$$

Or upon rearrangement, that we cannot simultaneously have

$$\sum_{i,j,k} kn^k_{ij} = 4r - 2d - 2g - (4 + \varepsilon) > \frac{r + 2}{2},$$
$$d + g + r \equiv 1 + \varepsilon \text{ or } 3 + \varepsilon \mod 5.$$

But by assumption (and because $\delta(n) = 4 + \varepsilon$), we cannot simultaneously have

$$\sum_{i,j,k} kn^k_{ij} = 4r - 2d - 2g - (4 + \varepsilon) > \frac{r}{2},$$
$$d + g + r \equiv 1 + \varepsilon \text{ or } 3 + \varepsilon \mod 5. \qquad \square$$

PROPOSITION A.31. *Suppose that Theorem A.18 holds for all $(d', g', r'; n')$ where either $d' < d$, or $d' = d$ and $\sum_{i,j,k} (n')^k_{ij} < \sum_{i,j,k} n^k_{ij}$. Then Theorem A.18 holds for $(d, g, r; n)$ if $\sum_{i,j,k} (i + j) \cdot n^k_{ij} = 2$.*

PROOF. By Theorem A.30, we may reduce to the case where

$$\frac{4(d+g)-4r}{2r-5} > \sum_{i,j,k} (i+j) \cdot n_{ij}^k = 2;$$

or upon rearrangement,

$$d+g > 2r-2-\frac{1}{2} \quad \Rightarrow \quad d+g \geq 2r-2.$$

If

$$\sum_{i,j,k} ((r-2)i + (r-3)j - k) \cdot n_{ij}^k \leq 2d+2g-3r,$$

then by Theorem A.23, it suffices to consider the case $\sum_{i,j,k} kn_{ij}^k < r-2$. In this case, Theorem 10.2 implies the desired result: We are reduced to showing $(d-1, g-1, r; n')$ is good, where crucially we do *not* have

$$\sum_{i,j,k} i(n')_{ij}^k = \sum_{i,j,k} j(n')_{ij}^k = 1.$$

In particular, our inductive hypothesis implies the desired result so long as

$$d+g-2 = (d-1)+(g-1) \geq \frac{7r-7}{4}.$$

This inequality holds since for $r \geq 12$,

$$d+g-2 \geq 2r-4 \geq \frac{7r-7}{4}.$$

On the other hand, if

$$\sum_{i,j,k} ((r-2)i + (r-3)j - k) \cdot n_{ij}^k \geq 2d+2g-3r+2,$$

then by Theorem A.24, we may also assume $\sum_{i,j,k} kn_{ij}^k < r-2$. In this case, we claim the desired result follows from Theorem 10.5. Indeed, we are reduced to showing $(d-1, g, r-1; n')$ is good, where we again crucially do *not* have

$$\sum_{i,j,k} i(n')_{ij}^k = \sum_{i,j,k} j(n')_{ij}^k = 1.$$

In particular, our inductive hypothesis implies the desired result so long as

$$d+g-1 = (d-1)+g \geq \frac{7(r-1)-7}{4} = \frac{7r-14}{4}.$$

This inequality holds since for $r \geq 12$,

$$d+g-1 \geq 2r-3 \geq \frac{7r-14}{4}.$$

It thus remains to consider the case where

$$\sum_{i,j,k} ((r-2)i + (r-3)j - k) \cdot n_{ij}^k = 2d+2g-3r+1.$$

By assumption, we have either $\sum_{i,j,k} in_{ij}^k = 2$ and $\sum_{i,j,k} jn_{ij}^k = 0$, or $\sum_{i,j,k} in_{ij}^k = \sum_{i,j,k} jn_{ij}^k = 1$.

First we consider the cases where either $\sum_{i,j,k} in_{ij}^k = 2$ and $\sum_{i,j,k} jn_{ij}^k = 0$, or $\sum_{i,j,k} kn_{ij}^k < r-3$. In either of these cases, we claim the desired result follows from Theorem 10.6. Indeed, we are reduced to showing $(d-2, g-1, r-1; n')$ is good, where we again crucially do *not* have

$$\sum_{i,j,k} i(n')_{ij}^k = \sum_{i,j,k} j(n')_{ij}^k = 1.$$

In particular, our inductive hypothesis implies the desired result so long as

$$d+g-3 = (d-2)+(g-1) \geq \frac{7(r-1)-7}{4} = \frac{7r-14}{4}.$$

This inequality holds since for $r \geq 12$,

$$d+g-3 \geq 2r-5 \geq \frac{7r-14}{4}.$$

Thus, it remains to consider the case

$$\sum_{i,j,k} in_{ij}^k = \sum_{i,j,k} jn_{ij}^k = 1$$

and $\sum_{i,j,k} kn_{ij}^k \in \{r-2, r-3\}$. In this case, we have

$$(r-2)+(r-3)-\sum_{i,j,k} kn_{ij}^k = \sum_{i,j,k}((r-2)i+(r-3)j-k)\cdot n_{ij}^k = 2d+2g-3r+1;$$

or upon rearrangement,

$$\sum_{i,j,k} kn_{ij}^k = 5r-2d-2g-6 \equiv r \mod 2.$$

It follows that in fact,

$$\sum_{i,j,k} kn_{ij}^k = r-2,$$

and that

$$r-2 = 5r-2d-2g-6 \quad \Rightarrow \quad d+g = 2r-2.$$

But this case is excluded by assumption. □

PROPOSITION A.32. *Suppose that Theorem A.18 holds for all $(d', g', r'; n')$ where either $d' < d$, or $d' = d$ and $\sum_{i,j,k}(n')_{ij}^k < \sum_{i,j,k} n_{ij}^k$. Then Theorem A.18 holds for $(d, g, r; n)$ if $\sum_{i,j,k}(i+j)\cdot n_{ij}^k = 1$.*

PROOF. Consider first the case when

$$\sum_{i,j,k}((r-2)i+(r-3)j-k)\cdot n_{ij}^k \neq 2d+2g-3r+1 \quad \text{and} \quad \sum_{i,j,k} kn_{ij}^k \geq r-3.$$

If in addition

$$\sum_{i,j,k}((r-2)i+(r-3)j-k)\cdot n_{ij}^k < 2d+2g-3r+1,$$

then Theorem A.23 implies the desired result. Similarly, if

$$\sum_{i,j,k}((r-2)i+(r-3)j-k)\cdot n_{ij}^k > 2d+2g-3r+1 \quad \text{and} \quad \sum_{i,j,k} kn_{ij}^k = r-2,$$

then Theorem A.24 implies the desired result. In this case we may thus assume

$$\sum_{i,j,k}((r-2)i+(r-3)j-k)\cdot n_{ij}^k > 2d+2g-3r+1 \quad \text{and} \quad \sum_{i,j,k} kn_{ij}^k = r-3.$$

Applying Theorem 10.5, it suffices to show $(d-1, g, r-1; n')$ satisfies our inductive hypothesis, where

$$(n')^k_{ij} = \begin{cases} \sum_\ell n^k_{\ell i} & \text{if } j = 0 \text{ and } (i, j; k) \neq (0, 0; 0); \\ 0 & \text{else.} \end{cases}$$

For this we first note that $\sum_{i,j,k} j(n')^k_{ij} = 0$; since $\delta(n') = 2$, our problem is thus reduced to showing that we do not simultaneously have

$$r - 3 = \sum_{i,j,k} kn^k_{ij} = \sum_{i,j,k} k(n')^k_{ij} = 4(r-1) - 2(d-1) - 2g - 2,$$
$$(d-1) + g + (r-1) \equiv 4 \text{ or } 1 \mod 5.$$

Reducing the first equation mod 5 and rearranging, it suffices to show that we do not simultaneously have

$$d + g + r \equiv 2 \mod 5$$
$$d + g + r \equiv 1 \text{ or } 3 \mod 5,$$

which is clear.

Next, we consider the case when

$$\sum_{i,j,k} ((r-2)i + (r-3)j - k) \cdot n^k_{ij} \leq 2d + 2g - 3r - 1 \quad \text{and} \quad \sum_{i,j,k} kn^k_{ij} < r - 3.$$

By Theorem 10.2 it is thus sufficient to show $(d-1, g-1, r; n')$ satisfies our inductive hypothesis, where

$$(n')^k_{ij} = \begin{cases} n^k_{ij} & \text{if } (i, j; k) \neq (1, 1; 1); \\ n^1_{11} + 1 & \text{if } (i, j; k) = (1, 1; 1). \end{cases}$$

For this, we first note that $\sum_{i,j,k} k(n')^k_{ij} < r - 2$; since $\delta(n') = 5$, our problem is thus reduced to showing

$$1 + \sum_{i,j,k} kn^k_{ij} = \sum_{i,j,k} k(n')^k_{ij} \neq 4r - 2(d-1) - 2(g-1) - 5;$$

or upon rearrangement, that

$$\sum_{i,j,k} kn^k_{ij} \neq 4r - 2d - 2g - 2.$$

But we have

$$(r-2) - \sum_{i,j,k} kn^k_{ij} = \sum_{i,j,k} ((r-2)i + (r-3)j - k) \cdot n^k_{ij} \leq 2d + 2g - 3r - 1;$$

which upon rearrangement gives

$$\sum_{i,j,k} kn^k_{ij} \geq 4r - 2d - 2g - 1,$$

completing the proof in this case.

Next, we consider the case when

$$\sum_{i,j,k} ((r-2)i + (r-3)j - k) \cdot n^k_{ij} \geq 2d + 2g - 3r + 3 \quad \text{and} \quad \sum_{i,j,k} kn^k_{ij} < r - 3.$$

Applying Theorem 10.5, it suffices to show $(d-1, g, r-1; n')$ satisfies our inductive hypothesis, where

$$(n')_{ij}^k = \begin{cases} \sum_\ell n_{\ell i}^k & \text{if } j = 0 \text{ and } (i,j;k) \neq (0,0;0); \\ 0 & \text{else.} \end{cases}$$

For this we first note that $\sum_{i,j,k} j(n')_{ij}^k = 0$; since $\delta(n') = 2$, our problem is thus reduced to showing

$$\sum_{i,j,k} kn_{ij}^k = \sum_{i,j,k} k(n')_{ij}^k \neq 4(r-1) - 2(d-1) - 2g - 2 = 4r - 2d - 2g - 4.$$

But we have

$$(r-2) - \sum_{i,j,k} kn_{ij}^k = \sum_{i,j,k} ((r-2)i + (r-3)j - k) \cdot n_{ij}^k \geq 2d + 2g - 3r + 3;$$

which upon rearrangement gives

$$\sum_{i,j,k} kn_{ij}^k \leq 4r - 2d - 2g - 5,$$

completing the proof in this case.

Next, we consider the case when

$$2d+2g-3r \leq \sum_{i,j,k} ((r-2)i+(r-3)j-k) \cdot n_{ij}^k \leq 2d+2g-3r+2 \quad \text{and} \quad \sum_{i,j,k} kn_{ij}^k < r-3.$$

In this case, we seek to apply Theorem 10.6. For this, it suffices to show $(d-2, g-1, r-1; n')$ is good, where

$$(n')_{ij}^k = \begin{cases} \sum_\ell n_{\ell i}^k & \text{if } j = 0 \text{ and } (i,j;k) \notin \{(0,0;0),(2,0,1)\}; \\ 1 + \sum_\ell n_{\ell i}^k & \text{if } (i,j;k) = (2,0,1); \\ 0 & \text{else.} \end{cases}$$

Since

$$\sum_{i,j,k} i(n')_{ij}^k = 2 + \sum_{i,j,k} jn_{ij}^k = 2 \quad \text{and} \quad \sum_{i,j,k} j(n')_{ij}^k = 0,$$

we have $\delta(n') = 4$, and it suffices to show that we do not simultaneously have

$$1 + \sum_{i,j,k} kn_{ij}^k = \sum_{i,j,k} k(n')_{ij}^k = 4(r-1) - 2(d-2) - 2(g-1) - 4 > \frac{(r-1)+3}{2},$$
$$(d-2) + (g-1) + (r-1) \equiv 1 \text{ or } 3 \mod 5.$$

Or, upon rearrangement, that we do not simultaneously have

$$\sum_{i,j,k} kn_{ij}^k = 4r - 2d - 2g - 3 > \frac{r}{2},$$
$$d + g + r \equiv 0 \text{ or } 2 \mod 5.$$

But this is precisely the assumption of Theorem A.18 (we have $\delta(n) = 3$).

It thus remains to consider the case when

$$\sum_{i,j,k} ((r-2)i + (r-3)j - k) \cdot n_{ij}^k = 2d + 2g - 3r + 1 \quad \text{and} \quad \sum_{i,j,k} kn_{ij}^k \in \{r-3, r-2\}.$$

Our first equation gives

$$(r-2) - \sum_{i,j,k} kn_{ij}^k = 2d + 2g - 3r + 1 \quad \Rightarrow \quad \sum_{i,j,k} kn_{ij}^k = 4r - 2d - 2g - 3.$$

This in addition implies

$$4r - 2d - 2g - 3 = \sum_{i,j,k} kn_{ij}^k = r - 3 \text{ or } r - 2.$$

Reducing mod 5 and rearranging, we obtain

$$r + d + g \equiv 0 \text{ or } 2 \mod 5.$$

But this case is excluded by assumption, since $\delta(n) = 3$ and $\sum_{i,j,k} kn_{ij}^k \geq r - 3 > r/2$. □

PROPOSITION A.33. *Suppose that Theorem A.18 holds for all $(d', g', r'; n')$ where either $d' < d$, or $d' = d$ and $\sum_{i,j,k} (n')_{ij}^k < \sum_{i,j,k} n_{ij}^k$. Then Theorem A.18 holds for $(d, g, r; n)$ if $\sum_{i,j,k} (i + j) \cdot n_{ij}^k = 0$.*

PROOF. Consider first the case when

$$\sum_{i,j,k} ((r-2)i + (r-3)j - k) \cdot n_{ij}^k \leq 2d + 2g - 3r.$$

By Theorem A.23, it suffices to consider the case $\sum_{i,j,k} kn_{ij}^k < r - 3$. By Theorem 10.2 it is thus sufficient to show $(d-1, g-1, r; n')$ satisfies our inductive hypothesis, where

$$(n')_{ij}^k = \begin{cases} n_{ij}^k & \text{if } (i,j;k) \neq (1,1;1); \\ n_{11}^1 + 1 & \text{if } (i,j;k) = (1,1;1). \end{cases}$$

For this, we first note that $\sum_{i,j,k} k(n')_{ij}^k < r - 2$; since $\delta(n') = 5$, our problem is thus reduced to showing that we do not simultaneously have

$$1 + \sum_{i,j,k} kn_{ij}^k = \sum_{i,j,k} k(n')_{ij}^k = 4r - 2(d-1) - 2(g-1) - 5 > \frac{r+3}{2},$$
$$(d-1) + (g-1) + r \equiv 5 + 2 \text{ or } 5 + 4 \mod 5.$$

Or equivalently, that we do not simultaneously have

$$\sum_{i,j,k} kn_{ij}^k = 4r - 2d - 2g - 2 > \frac{r+1}{2},$$
$$d + g + r \equiv 4 \text{ or } 1 \mod 5.$$

But by assumption (since $\delta(n) = 2$), we do not simultaneously have

$$\sum_{i,j,k} kn_{ij}^k = 4r - 2d - 2g - 2 > \frac{r}{2},$$
$$d + g + r \equiv 4 \text{ or } 1 \mod 5.$$

Next, we consider the case when

$$\sum_{i,j,k} ((r-2)i + (r-3)j - k) \cdot n_{ij}^k \geq 2d + 2g - 3r + 1. \tag{A.6}$$

In this case, we first of all claim that the inequalities of Theorem 10.10 and Theorem 10.7 are satisfied. To check this, we apply Theorem A.19, which reduces our claim to verifying the inequality

$$\sum_{i,j,k} kn_{ij}^k \leq \frac{3r^2 - 3r - 4}{2r - 4} - \frac{r - 5}{2r - 4}(d + g).$$

On the other hand, our assumption implies

$$\begin{aligned} -\sum_{i,j,k} kn_{ij}^k &= \sum_{i,j,k} ((r-2)i + (r-3)j - k) \cdot n_{ij}^k \\ &\geq 2d + 2g - 3r + 1 \quad \Rightarrow \quad d + g \leq \frac{3r - 1 - \sum_{i,j,k} kn_{ij}^k}{2}. \end{aligned}$$

We are thus reduced to showing

$$\sum_{i,j,k} kn_{ij}^k \leq \frac{3r^2 - 3r - 4}{2r - 4} - \frac{r - 5}{2r - 4} \cdot \frac{3r - 1 - \sum_{i,j,k} kn_{ij}^k}{2}.$$

Or, upon rearrangement, that

$$\sum_{i,j,k} kn_{ij}^k \leq \frac{3r + 13}{3}.$$

For this is it sufficient to note that

$$r - 2 \leq \frac{3r + 13}{3}.$$

Now if $\sum_{i,j,k} n_{ij}^k > 0$, we may iteratively apply Theorem 10.10 — noting that if n' is as in Theorem 10.10, then $(d, g, r; n')$ also satisfies the inequality of Theorem 10.10 — to reduce to cases where $\sum_{i,j,k} n_{ij}^k = 0$ (which are good by Condition 3 of Theorem A.21). Similarly, if $\sum_{i,j,k} n_{ij}^k = 0$ and $d > g + r$, we may apply Theorem 10.7, again applying Condition 3 of Theorem A.21 to check the inductive hypothesis. It therefore remains to consider the case where $d = g + r$ and $\sum_{i,j,k} n_{ij}^k = 0$.

If the inequality Eq. (A.6) is strict, then the desired result follows from Theorem 10.5: The n' appearing in Theorem 10.5 satisfies

$$\sum_{i,j,k} k(n')_{ij}^k = 0.$$

We may thus suppose additionally that Eq. (A.6) is an equality. But in this case, we additionally have

$$2d + 2g - 3r + 1 = \sum_{i,j,k} ((r-2)i + (r-3)j - k) \cdot n_{ij}^k = 0.$$

Using the above equation together with $d = g + r$ to solve for d and r in terms of g, we obtain

$$d = 5g + 1 \quad \text{and} \quad r = 4g + 1.$$

We are thus done by Theorem 11.3. □

APPENDIX B

Code for Chapter 4

In this aapendix, we give python code to do the finite computations described in Chapter 4.

```
class Point:
  def __init__(self, i, j, k):
    self.i = i
    self.j = j
    self.k = k

  def __repr__(self):
    return '(' + str(self.i) + ', ' + str(self.j) + '; ' + str(self.k)
        + ')'

  def as_tuple(self):
    return (self.i, self.j, self.k)

  def __hash__(self):
    return hash(self.as_tuple())

  def __eq__(self, other):
    return self.as_tuple() == other.as_tuple()

  def __ne__(self, other):
    return not (self == other)

# Allowed types of markings for points:
P111 = Point(1, 1, 1)
P201 = Point(2, 0, 1)
P102 = Point(1, 0, 2)
P110 = Point(1, 1, 0)
P101 = Point(1, 0, 1)
P200 = Point(2, 0, 0)
P002 = Point(0, 0, 2)
P100 = Point(1, 0, 0)
P001 = Point(0, 0, 1)
P000 = Point(0, 0, 0)

BLANK = P000 # Blank point (to eliminate).
POINTS = [P111, P201, P102, P110, P101, P200, P002, P100, P001]
NO_MARKINGS = {P:0 for P in POINTS}

class Curve:
  def __init__(self, d, g, r, m = NO_MARKINGS.copy()):
```

```
    self.d = d
    self.g = g
    self.r = r

    self.m = m

    for i in m.keys():
      if i not in POINTS:
        raise ValueError

    self.nm = sum([self.m[P] for P in POINTS]) # Number of marked
        points.

    self.I = sum([self.m[P] * P.i for P in POINTS]) # \sum i n_{ij}^k
    self.J = sum([self.m[P] * P.j for P in POINTS]) # \sum j n_{ij}^k
    self.K = sum([self.m[P] * P.k for P in POINTS]) # \sum k n_{ij}^k

    self.lhs = (r - 2) * self.I + (r - 3) * self.J - self.K # \sum [(r
        - 2)i + (r - 3)j - k] n_{ij}^k

    if self.K > r - 2:
      raise ValueError, "K is too big."

  def as_tuple(self):
    return (self.d, self.g, self.r, tuple([self.m[P] for P in POINTS])
        )

  def __hash__(self):
    return hash(self.as_tuple())

  def __eq__(self, other):
    return self.as_tuple() == other.as_tuple()

  def __ne__(self, other):
    return not (self == other)

  def __repr__(self):
    out = 'Curve of degree ' + str(self.d) + ' and genus ' + str(self.
        g) + ' in P^' + str(self.r)
    if self.m != NO_MARKINGS:
      out += ', with marked points:'
      for P in POINTS:
        if self.m[P]:
          out += '\n    ' + str(P) + ' x ' + str(self.m[P])
    return out + str('.')

  def delete(self, P, n = 1):
    mprime = self.m.copy()
    if mprime[P] < n:
      raise ValueError, "Cannot delete a point that does not exist."
    mprime[P] -= n
    return Curve(self.d, self.g, self.r, mprime)
```

```
  def add(self, P, n = 1):
    mprime = self.m.copy()
    mprime[P] += n
    return Curve(self.d, self.g, self.r, mprime)

  def replace(self, f):
    mprime = NO_MARKINGS.copy()
    for P in POINTS:
      if f(P) != BLANK:
        mprime[f(P)] += self.m[P]
    return Curve(self.d, self.g, self.r, mprime)

  def lower_d(self, n = 1):
    return Curve(self.d - n, self.g, self.r, self.m)

  def lower_g(self, n = 1):
    return Curve(self.d, self.g - n, self.r, self.m)

  def lower_r(self, n = 1):
    return Curve(self.d, self.g, self.r - n, self.m)

def partition(ijk, types = POINTS):
  if len(types) == 0:
    if ijk == [0, 0, 0]:
      yield {}
    return

  t = types[0].as_tuple()
  limits = []

  for l in (0, 1, 2):
    if t[l] != 0:
      limits.append(ijk[l] / t[l])

  for n in xrange(1 + min(limits)):
    for P in partition([ijk[l] - n * t[l] for l in (0, 1, 2)], types
        [1:]):
      P[types[0]] = n
      yield P.copy()
  return

def all_curves(d, g, r):
  bound = 2 * d + 2 * g - r - 2
  for k in xrange(r - 1):
    for i in xrange(1 + bound + k):
      for j in xrange(1 + i):
        if (r - 2) * i + (r - 3) * j - k <= bound:
          for m in partition([i, j, k]):
            yield Curve(d, g, r, m)

GOOD = {}
def good(C):
  if C in GOOD:
```

```
    return GOOD[C]

  if C.lhs > 2 * C.d + 2 * C.g - C.r - 2:
    return False

  if C.g == 0:
    return True

  if C.r == 2:
    return True

  if C.r == 3:
    if C.K == 0:
      if (C.J != 0) or (C.I != 2 * C.d + 2 * C.g - 14):
        return True
    else:
      if C.I != 2 * C.d + 2 * C.g - 9:
        return True

  if (C.nm == 0) and (C.d == 5 * C.g + 1) and (C.r == 4 * C.g + 1):
    return True

  if C == Curve(8, 3, 5).add(P101, 2):
    return True

  if (C.nm == 0) and (C.r == 5) and (C.g >= 2):
    if good(C.lower_d(2).lower_g(2).add(P101, 2)):
      GOOD[C] = True; return True

  if (C.r - 1) * C.nm - C.I - 2 * C.J - C.K <= (C.r + 1) * C.d - (2 *
      C.r - 4) * C.g - 2:
    for P in POINTS:
      if (P.j == P.k == 0) and (C.m[P] > 0):
        if C.K < C.r - 2:
          if good(C.delete(P)) and good(C.replace(lambda P : Point(P.j
              , 0, P.k)).lower_r().lower_d()):
            GOOD[C] = True; return True
        else:
          if good(C.delete(P)) and good(C.replace(lambda P : Point(P.j
              , 0, 0)).lower_r().lower_d()):
            GOOD[C] = True; return True

      if (P.i == P.j == 0) and (C.m[P] > 0):
        if good(C.delete(P)) and good(C.replace(lambda P : Point(P.i,
            P.j, 0))):
          GOOD[C] = True; return True

    if C.d > C.g + C.r:
      if (C.r != 3) or (C.J == 0):
        if C.K < C.r - 2:
          if good(C.lower_d()) and good(C.lower_d().lower_r()):
            GOOD[C] = True; return True
        else:
```

```
        if good(C.lower_d()) and good(C.replace(lambda P : Point(P.i
            , P.j, 0)).lower_d().lower_r()):
          GOOD[C] = True; return True

  if C.K < C.r - 2:
    if good(C.add(P111).lower_d().lower_g()):
      GOOD[C] = True; return True
  else:
    if good(C.replace(lambda P : Point(P.i, P.j, 0)).add(P110).lower_d
        ().lower_g()):
      GOOD[C] = True; return True

  if (C.r != 3):
    if C.K == C.r - 3:
      if good(C.replace(lambda P : Point(P.i, P.j, 0)).add(P101).
          lower_d().lower_g()):
        GOOD[C] = True; return True
    if C.K == C.r - 2:
      if good(C.replace(lambda P : Point(P.i, P.j, 0)).add(P102).
          lower_d().lower_g()):
        GOOD[C] = True; return True

  if C.lhs >= 2 * C.d + 2 * C.g - 3 * C.r + 2:
    if C.K < C.r - 2:
      if good(C.replace(lambda P : Point(P.j, 0, P.k)).lower_d().
          lower_r()):
        GOOD[C] = True; return True
    else:
      if good(C.replace(lambda P : Point(P.j, 0, 0)).lower_d().lower_r
          ()):
        GOOD[C] = True; return True

  if (C.r != 3) and (2 * C.d + 2 * C.g - 4 * C.r + 3 <= C.lhs <= 2 * C
      .d + 2 * C.g - 2 * C.r - 1):
    if C.K < C.r - 3:
      if good(C.replace(lambda P : Point(P.j, 0, P.k)).add(P201).
          lower_d(2).lower_g().lower_r()):
        GOOD[C] = True; return True
    elif C.K == C.r - 3:
      if good(C.replace(lambda P : Point(P.j, 0, 0)).add(P200).lower_d
          (2).lower_g().lower_r()):
        GOOD[C] = True; return True
    else:
      if good(C.replace(lambda P : Point(P.j, 0, 0)).add(P201).lower_d
          (2).lower_g().lower_r()):
        GOOD[C] = True; return True

  GOOD[C] = False; return False

def check(r, dg_max):
  largest_dg = 0
  for d in xrange(r, dg_max + 1):
    for g in xrange(min(d - r, dg_max - d) + 1):
```

```
    for C in all_curves(d, g, r):
      if not good(C):
        largest_dg = max(largest_dg, d + g)
        if C.m == NO_MARKINGS:
          print 'Potential counterexample:', C
  if largest_dg != 0:
    print 'Largest potentially non-excellent d + g =', largest_dg

class Point:
  def __init__(self, i, j, k):
    self.i = i
    self.j = j
    self.k = k

  def __repr__(self):
    return '(' + str(self.i) + ', ' + str(self.j) + '; ' + str(self.k)
        + ')'

  def as_tuple(self):
    return (self.i, self.j, self.k)

  def __hash__(self):
    return hash(self.as_tuple())

  def __eq__(self, other):
    return self.as_tuple() == other.as_tuple()

  def __ne__(self, other):
    return not (self == other)

# Allowed types of markings for points:
P111 = Point(1, 1, 1)
P201 = Point(2, 0, 1)
P102 = Point(1, 0, 2)
P110 = Point(1, 1, 0)
P101 = Point(1, 0, 1)
P200 = Point(2, 0, 0)
P002 = Point(0, 0, 2)
P100 = Point(1, 0, 0)
P001 = Point(0, 0, 1)
P000 = Point(0, 0, 0)

BLANK = P000 # Blank point (to eliminate).
POINTS = [P111, P201, P102, P110, P101, P200, P002, P100, P001]
NO_MARKINGS = {P:0 for P in POINTS}

class Curve:
  def __init__(self, d, g, r, m = NO_MARKINGS.copy()):
    self.d = d
    self.g = g
    self.r = r

    self.m = m
```

```
    for i in m.keys():
      if i not in POINTS:
        raise ValueError

    self.nm = sum([self.m[P] for P in POINTS]) # Number of marked
        points.

    self.I = sum([self.m[P] * P.i for P in POINTS]) # \sum i n_{ij}^k
    self.J = sum([self.m[P] * P.j for P in POINTS]) # \sum j n_{ij}^k
    self.K = sum([self.m[P] * P.k for P in POINTS]) # \sum k n_{ij}^k

    self.lhs = (r - 2) * self.I + (r - 3) * self.J - self.K # \sum [(r
        - 2)i + (r - 3)j - k] n_{ij}^k

    if self.K > r - 2:
      raise ValueError, "K is too big."

  def as_tuple(self):
    return (self.d, self.g, self.r, tuple([self.m[P] for P in POINTS])
        )

  def __hash__(self):
    return hash(self.as_tuple())

  def __eq__(self, other):
    return self.as_tuple() == other.as_tuple()

  def __ne__(self, other):
    return not (self == other)

  def __repr__(self):
    out = 'Curve of degree ' + str(self.d) + ' and genus ' + str(self.
        g) + ' in P^' + str(self.r)
    if self.m != NO_MARKINGS:
      out += ', with marked points:'
      for P in POINTS:
        if self.m[P]:
          out += '\n    ' + str(P) + ' x ' + str(self.m[P])
    return out + str('.')

  def delete(self, P, n = 1):
    mprime = self.m.copy()
    if mprime[P] < n:
      raise ValueError, "Cannot delete a point that does not exist."
    mprime[P] -= n
    return Curve(self.d, self.g, self.r, mprime)

  def add(self, P, n = 1):
    mprime = self.m.copy()
    mprime[P] += n
    return Curve(self.d, self.g, self.r, mprime)
```

```
  def replace(self, f):
    mprime = NO_MARKINGS.copy()
    for P in POINTS:
      if f(P) != BLANK:
        mprime[f(P)] += self.m[P]
    return Curve(self.d, self.g, self.r, mprime)

  def lower_d(self, n = 1):
    return Curve(self.d - n, self.g, self.r, self.m)

  def lower_g(self, n = 1):
    return Curve(self.d, self.g - n, self.r, self.m)

  def lower_r(self, n = 1):
    return Curve(self.d, self.g, self.r - n, self.m)

def partition(ijk, types = POINTS):
  if len(types) == 0:
    if ijk == [0, 0, 0]:
      yield {}
    return

  t = types[0].as_tuple()
  limits = []

  for l in (0, 1, 2):
    if t[l] != 0:
      limits.append(ijk[l] / t[l])

  for n in xrange(1 + min(limits)):
    for P in partition([ijk[l] - n * t[l] for l in (0, 1, 2)], types
        [1:]):
      P[types[0]] = n
      yield P.copy()
  return

def all_curves(d, g, r):
  bound = 2 * d + 2 * g - r - 2
  for k in xrange(r - 1):
    for i in xrange(1 + bound + k):
      for j in xrange(1 + i):
        if (r - 2) * i + (r - 3) * j - k <= bound:
          for m in partition([i, j, k]):
            yield Curve(d, g, r, m)

GOOD = {}
def good(C):
  if C in GOOD:
    return GOOD[C]

  if C.lhs > 2 * C.d + 2 * C.g - C.r - 2:
    return False
```

```
if C.g == 0:
  return True

if C.r == 2:
  return True

if C.r == 3:
  if C.K == 0:
    if (C.J != 0) or (C.I != 2 * C.d + 2 * C.g - 14):
      return True
  else:
    if C.I != 2 * C.d + 2 * C.g - 9:
      return True

if (C.nm == 0) and (C.d == 5 * C.g + 1) and (C.r == 4 * C.g + 1):
  return True

if C == Curve(8, 3, 5).add(P101, 2):
  return True

if (C.nm == 0) and (C.r == 5) and (C.g >= 2):
  if good(C.lower_d(2).lower_g(2).add(P101, 2)):
    GOOD[C] = True; return True

if (C.r - 1) * C.nm - C.I - 2 * C.J - C.K <= (C.r + 1) * C.d - (2 *
    C.r - 4) * C.g - 2:
  for P in POINTS:
    if (P.j == P.k == 0) and (C.m[P] > 0):
      if C.K < C.r - 2:
        if good(C.delete(P)) and good(C.replace(lambda P : Point(P.j
            , 0, P.k)).lower_r().lower_d()):
          GOOD[C] = True; return True
      else:
        if good(C.delete(P)) and good(C.replace(lambda P : Point(P.j
            , 0, 0)).lower_r().lower_d()):
          GOOD[C] = True; return True

    if (P.i == P.j == 0) and (C.m[P] > 0):
      if good(C.delete(P)) and good(C.replace(lambda P : Point(P.i,
          P.j, 0))):
        GOOD[C] = True; return True

  if C.d > C.g + C.r:
    if (C.r != 3) or (C.J == 0):
      if C.K < C.r - 2:
        if good(C.lower_d()) and good(C.lower_d().lower_r()):
          GOOD[C] = True; return True
      else:
        if good(C.lower_d()) and good(C.replace(lambda P : Point(P.i
            , P.j, 0)).lower_d().lower_r()):
          GOOD[C] = True; return True

if C.K < C.r - 2:
```

```
    if good(C.add(P111).lower_d().lower_g()):
      GOOD[C] = True; return True
  else:
    if good(C.replace(lambda P : Point(P.i, P.j, 0)).add(P110).lower_d
        ().lower_g()):
      GOOD[C] = True; return True

  if (C.r != 3):
    if C.K == C.r - 3:
      if good(C.replace(lambda P : Point(P.i, P.j, 0)).add(P101).
          lower_d().lower_g()):
        GOOD[C] = True; return True
    if C.K == C.r - 2:
      if good(C.replace(lambda P : Point(P.i, P.j, 0)).add(P102).
          lower_d().lower_g()):
        GOOD[C] = True; return True

  if C.lhs >= 2 * C.d + 2 * C.g - 3 * C.r + 2:
    if C.K < C.r - 2:
      if good(C.replace(lambda P : Point(P.j, 0, P.k)).lower_d().
          lower_r()):
        GOOD[C] = True; return True
    else:
      if good(C.replace(lambda P : Point(P.j, 0, 0)).lower_d().lower_r
          ()):
        GOOD[C] = True; return True

  if (C.r != 3) and (2 * C.d + 2 * C.g - 4 * C.r + 3 <= C.lhs <= 2 * C
      .d + 2 * C.g - 2 * C.r - 1):
    if C.K < C.r - 3:
      if good(C.replace(lambda P : Point(P.j, 0, P.k)).add(P201).
          lower_d(2).lower_g().lower_r()):
        GOOD[C] = True; return True
    elif C.K == C.r - 3:
      if good(C.replace(lambda P : Point(P.j, 0, 0)).add(P200).lower_d
          (2).lower_g().lower_r()):
        GOOD[C] = True; return True
    else:
      if good(C.replace(lambda P : Point(P.j, 0, 0)).add(P201).lower_d
          (2).lower_g().lower_r()):
        GOOD[C] = True; return True

  GOOD[C] = False; return False

def check(r, dg_max):
  largest_dg = 0
  for d in xrange(r, dg_max + 1):
    for g in xrange(min(d - r, dg_max - d) + 1):
      for C in all_curves(d, g, r):
        if not good(C):
          largest_dg = max(largest_dg, d + g)
          if C.m == NO_MARKINGS:
            print 'Potential counterexample:', C
```

```
    if largest_dg != 0:
      print 'Largest potentially non-excellent d + g =', largest_dg
```

The output is as follows:

```
>>> check(4, 16)
Potential counterexample: Curve of degree 6 and genus 2 in P^4.
Largest potentially non-excellent d + g = 10
>>> check(5, 15)
Potential counterexample: Curve of degree 7 and genus 2 in P^5.
Largest potentially non-excellent d + g = 13
>>> check(6, 16)
Largest potentially non-excellent d + g = 12
>>> check(7, 15)
Largest potentially non-excellent d + g = 13
>>> check(8, 16)
>>> for r in xrange(9, 12):
...     check(r, 2 * r - 2)
...
>>>
```


```
>>> check(4, 16)
Potential counterexample: Curve of degree 6 and genus 2 in P^4.
Largest potentially non-excellent d + g = 10
>>> check(5, 15)
Potential counterexample: Curve of degree 7 and genus 2 in P^5.
Largest potentially non-excellent d + g = 13
>>> check(6, 16)
Largest potentially non-excellent d + g = 12
>>> check(7, 15)
Largest potentially non-excellent d + g = 13
>>> check(8, 16)
>>> for r in xrange(9, 12):
...     check(r, 2 * r - 2)
...
>>>
```


Acknowledgments

The authors would like to thank Joe Harris for introducing us to each other, and for his guidence throughout this research. We are also grateful to Bjorn Poonen for many helful conversations. We would like to thank the members of the Harvard and MIT mathematics departments, in particular Gabriel Bujokas, Francesco Cavazzani, Anand Patel, Alexander Perry, and Eric Riedl; and to thank the anonymous referee for helpful suggestions. I, the first author would like to acknowledge the generous support of the National Science Foundation, grant DMS-1308244, for "Nonlinear Analysis on Sympletic, Complex Manifolds, General Relativity, and Graphs." The second author would also like to acknowledge the generous support both of the Fannie and John Hertz Foundation, and of the Department of Defense (NDSEG fellowship).

Bibliography

[1] A. Atanasov, *Interpolation and vector bundles on curves*, Preprint, math.AG:1404.4892 (2014).

[2] M. Bôcher, *The theory of linear dependence*, Ann. of Math. (2) **2** (1900/01), no. 1-4, 81–96, DOI 10.2307/2007186. MR1503482

[3] I. Coskun, *Degenerations of surface scrolls and the Gromov-Witten invariants of Grassmannians*, J. Algebraic Geom. **15** (2006), no. 2, 223–284, DOI 10.1090/S1056-3911-06-00426-7. MR2199064

[4] L. Ein and R. Lazarsfeld, *Stability and restrictions of Picard bundles, with an application to the normal bundles of elliptic curves*, Complex projective geometry (Trieste, 1989/Bergen, 1989), London Math. Soc. Lecture Note Ser., vol. 179, Cambridge Univ. Press, Cambridge, 1992, pp. 149–156, DOI 10.1017/CBO9780511662652.011. MR1201380

[5] D. Eisenbud and J. Harris, *On varieties of minimal degree (a centennial account)*, Algebraic geometry, Bowdoin, 1985 (Brunswick, Maine, 1985), Proc. Sympos. Pure Math., vol. 46, Amer. Math. Soc., Providence, RI, 1987, pp. 3–13, DOI 10.1090/pspum/046.1/927946. MR927946

[6] R. Hartshorne and A. Hirschowitz, *Smoothing algebraic space curves*, Algebraic geometry, Sitges (Barcelona), 1983, Lecture Notes in Math., vol. 1124, Springer, Berlin, 1985, pp. 98–131, DOI 10.1007/BFb0074998. MR805332

[7] R. Hartshorne, *Algebraic geometry*, Springer-Verlag, New York-Heidelberg, 1977. Graduate Texts in Mathematics, No. 52. MR0463157

[8] D. Perrin, *Courbes passant par m points généraux de* $\mathbf{P}^3$ (French, with English summary), Mém. Soc. Math. France (N.S.) **28-29** (1987), 138. MR925737

[9] Z. Ran, *Normal bundles of rational curves in projective spaces*, Asian J. Math. **11** (2007), no. 4, 567–608, DOI 10.4310/AJM.2007.v11.n4.a3. MR2402939

[10] G. Sacchiero, *Normal bundles of rational curves in projective space* (Italian, with English summary), Ann. Univ. Ferrara Sez. VII (N.S.) **26** (1980), 33–40 (1981). MR608295

[11] J. Stevens, *On the number of points determining a canonical curve*, Nederl. Akad. Wetensch. Indag. Math. **51** (1989), no. 4, 485–494. MR1041502

Editorial Information

To be published in the *Memoirs*, a paper must be correct, new, nontrivial, and significant. Further, it must be well written and of interest to a substantial number of mathematicians. Piecemeal results, such as an inconclusive step toward an unproved major theorem or a minor variation on a known result, are in general not acceptable for publication.

Papers appearing in *Memoirs* are generally at least 80 and not more than 200 published pages in length. Papers less than 80 or more than 200 published pages require the approval of the Managing Editor of the Transactions/Memoirs Editorial Board. Published pages are the same size as those generated in the style files provided for $\mathcal{A}_{\mathcal{M}}\mathcal{S}$-LaTeX or $\mathcal{A}_{\mathcal{M}}\mathcal{S}$-TeX.

Information on the backlog for this journal can be found on the AMS website starting from `http://www.ams.org/memo`.

A Consent to Publish is required before we can begin processing your paper. After a paper is accepted for publication, the Providence office will send a Consent to Publish and Copyright Agreement to all authors of the paper. By submitting a paper to the *Memoirs*, authors certify that the results have not been submitted to nor are they under consideration for publication by another journal, conference proceedings, or similar publication.

Information for Authors

Memoirs is an author-prepared publication. Once formatted for print and on-line publication, articles will be published as is with the addition of AMS-prepared frontmatter and backmatter. Articles are not copyedited; however, confirmation copy will be sent to the authors.

Initial submission. The AMS uses Centralized Manuscript Processing for initial submissions. Authors should submit a PDF file using the Initial Manuscript Submission form found at `www.ams.org/submission/memo`, or send one copy of the manuscript to the following address: Centralized Manuscript Processing, MEMOIRS OF THE AMS, 201 Charles Street, Providence, RI 02904-2294 USA. If a paper copy is being forwarded to the AMS, indicate that it is for *Memoirs* and include the name of the corresponding author, contact information such as email address or mailing address, and the name of an appropriate Editor to review the paper (see the list of Editors below).

The paper must contain a *descriptive title* and an *abstract* that summarizes the article in language suitable for workers in the general field (algebra, analysis, etc.). The *descriptive title* should be short, but informative; useless or vague phrases such as "some remarks about" or "concerning" should be avoided. The *abstract* should be at least one complete sentence, and at most 300 words. Included with the footnotes to the paper should be the 2010 *Mathematics Subject Classification* representing the primary and secondary subjects of the article. The classifications are accessible from `www.ams.org/msc/`. The Mathematics Subject Classification footnote may be followed by a list of *key words and phrases* describing the subject matter of the article and taken from it. Journal abbreviations used in bibliographies are listed in the latest *Mathematical Reviews* annual index. The series abbreviations are also accessible from `www.ams.org/msnhtml/serials.pdf`. To help in preparing and verifying references, the AMS offers MR Lookup, a Reference Tool for Linking, at `www.ams.org/mrlookup/`.

Electronically prepared manuscripts. The AMS encourages electronically prepared manuscripts, with a strong preference for $\mathcal{A}_{\mathcal{M}}\mathcal{S}$-LaTeX. To this end, the Society has prepared $\mathcal{A}_{\mathcal{M}}\mathcal{S}$-LaTeX author packages for each AMS publication. Author packages include instructions for preparing electronic manuscripts, samples, and a style file that generates the particular design specifications of that publication series. Though $\mathcal{A}_{\mathcal{M}}\mathcal{S}$-LaTeX is the highly preferred format of TeX, author packages are also available in $\mathcal{A}_{\mathcal{M}}\mathcal{S}$-TeX.

Authors may retrieve an author package for *Memoirs of the AMS* from `www.ams.org/journals/memo/memoauthorpac.html`. The *AMS Author Handbook* is available in PDF format from the author package link. The author package can also be obtained free of charge by sending email to tech-support@ams.org or from the Publication Division, American Mathematical Society, 201 Charles St., Providence, RI 02904-2294, USA. When requesting an author package, please specify $\mathcal{AMS}$-LaTeX or $\mathcal{AMS}$-TeX and the publication in which your paper will appear. Please be sure to include your complete mailing address.

After acceptance. The source files for the final version of the electronic manuscript should be sent to the Providence office immediately after the paper has been accepted for publication. The author should also submit a PDF of the final version of the paper to the editor, who will forward a copy to the Providence office.

Accepted electronically prepared files can be submitted via the web at www.ams.org/submit-book-journal/, sent via FTP, or sent on CD to the Electronic Prepress Department, American Mathematical Society, 201 Charles Street, Providence, RI 02904-2294 USA. TeX source files and graphic files can be transferred over the Internet by FTP to the Internet node ftp.ams.org (130.44.1.100). When sending a manuscript electronically via CD, please be sure to include a message indicating that the paper is for the *Memoirs*.

Electronic graphics. Comprehensive instructions on preparing graphics are available at www.ams.org/authors/journals.html. A few of the major requirements are given here.

Submit files for graphics as EPS (Encapsulated PostScript) files. This includes graphics originated via a graphics application as well as scanned photographs or other computer-generated images. If this is not possible, TIFF files are acceptable as long as they can be opened in Adobe Photoshop or Illustrator.

Authors using graphics packages for the creation of electronic art should also avoid the use of any lines thinner than 0.5 points in width. Many graphics packages allow the user to specify a "hairline" for a very thin line. Hairlines often look acceptable when proofed on a typical laser printer. However, when produced on a high-resolution laser imagesetter, hairlines become nearly invisible and will be lost entirely in the final printing process.

Screens should be set to values between 15% and 85%. Screens which fall outside of this range are too light or too dark to print correctly. Variations of screens within a graphic should be no less than 10%.

Any graphics created in color will be rendered in grayscale for the printed version unless color printing is authorized by the Managing Editor and the Publisher. In general, color graphics will appear in color in the online version.

Inquiries. Any inquiries concerning a paper that has been accepted for publication should be sent to memo-query@ams.org or directly to the Electronic Prepress Department, American Mathematical Society, 201 Charles St., Providence, RI 02904-2294 USA.

Editors

This journal is designed particularly for long research papers, normally at least 80 pages in length, and groups of cognate papers in pure and applied mathematics. Papers intended for publication in the *Memoirs* should be addressed to one of the following editors. The AMS uses Centralized Manuscript Processing for initial submissions to AMS journals. Authors should follow instructions listed on the Initial Submission page found at www.ams.org/memo/memosubmit.html.

1. GEOMETRY, TOPOLOGY & LOGIC
   Coordinating Editor: Richard Canary, Department of Mathematics, University of Michigan, Ann Arbor, MI 48109-1043 USA; e-mail: canary@umich.edu

   Algebraic topology, Michael Hill, Department of Mathematics, University of California Los Angeles, Los Angeles, CA 90095 USA; e-mail: mikehill@math.ucla.edu

   Differential geometry, Chiu-Chu Melissa Liu, Department of Mathematics, Columbia University, New York, NY 10027 USA; e-mail: ccliu@math.columbia.edu

   Logic, Noam Greenberg, School of Mathematics and Statistics, Victoria University of Wellington, Wellington 6140, New Zealand; e-mail: greenberg@msor.vuw.ac.nz

   Low-dimensional topology and geometric structures, Richard Canary

2. ALGEBRA AND NUMBER THEORY
   Coordinating Editor: Henri Darmon, Department of Mathematics, McGill University, Montreal, Quebec H3A 0G4, Canada; e-mail: darmon@math.mcgill.ca

   Algebra, Michael Larsen, Department of Mathematics, Rawles Hall, Indiana University, 831 E 3rd St., Bloomington, IN 47405 USA; e-mail: mjlarsen@indiana.edu

   Algebraic geometry, Lucia Caporaso, Department of Mathematics and Physics, Roma Tre University, Largo San Leonardo Murialdo, I-00146 Rome, Italy; e-mail: LCedit@mat.uniroma3.it

   Arithmetic geometry, Ted C. Chinburg, Department of Mathematics, University of Pennsylvania, Philadelphia, PA 19104-6395 USA; e-mail: ted@math.upenn.edu

   Commutative algebra, Irena Peeva, Department of Mathematics, Cornell University, Ithaca, NY 14853 USA; e-mail: irena@math.cornell.edu

   Number theory, Henri Darmon

3. GEOMETRIC ANALYSIS & PDE
   Coordinating Editor: Tatiana Toro, Department of Mathematics, University of Washington, Box 354350, Seattle, WA 98195-4350 USA; e-mail: toro@uw.edu

   Geometric analysis, Tatiana Toro

   Harmonic analysis and partial differential equations, Monica Visan, Department of Mathematics, University of California Los Angeles, 520 Portola Plaza, Los Angeles, CA 90095 USA; e-mail: visan@math.ucla.edu

   Partial differential equations and functional analysis, Alexander A. Kiselev, Department of Mathematics, Duke University, 120 Science Drive, Rm 117 Physics Bldg., Durham, NC 27708 USA; e-mail: kiselev@math.duke.edu

   Real analysis and partial differential equations, Wilhelm Schlag, Department of Mathematics, The University of Chicago, 5734 South University Avenue, Chicago, IL 60637 USA; e-mail: schlag@math.uchicago.edu

4. ERGODIC THEORY, DYNAMICAL SYSTEMS & COMBINATORICS
   Coordinating Editor: Vitaly Bergelson, Department of Mathematics, Ohio State University, 231 W. 18th Avenue, Columbus, OH 43210 USA; e-mail: vitaly@math.ohio-state.edu

   Algebraic and enumerative combinatorics, Jim Haglund, Department of Mathematics, University of Pennsylvania, Philadelphia, PA 19104 USA; e-mail: jhaglund@math.upenn.edu

   Probability theory, Robin Pemantle, Department of Mathematics, University of Pennsylvania, 209 S. 33rd Street, Philadelphia, PA 19104 USA; e-mail: pemantle@math.upenn.edu

   Dynamical systems and ergodic theory, Ian Melbourne, Mathematics Institute, University of Warwick, Coventry CV4 7AL, United Kingdom; e-mail: I.Melbourne@warwick.ac.uk

   Ergodic theory and combinatorics, Vitaly Bergelson

5. ANALYSIS, LIE THEORY & PROBABILITY
   Coordinating Editor: Stefaan Vaes, Department of Mathematics, Katholieke Universiteit Leuven, Celestijnenlaan 200B, B-3001 Leuven, Belgium; e-mail: stefaan.vaes@wis.kuleuven.be

   Functional analysis and operator algebras, Stefaan Vaes

   Harmonic analysis and complex analysis, Malabika Pramanik, Department of Mathematics, University of British Columbia, 1984 Mathematics Road, Vancouver, British Columbia V6T 1Z2, Canada; e-mail: malabika@math.ubc.ca

   Langlands conjectures, Marie-France Vigneras, 8 Rue des Ecoles, 75005 Paris, France; e-mail: marie-france.vigneras@imj-prg.fr

   Probability and statistics, Patrick J. Fitzsimmons, Department of Mathematics, University of California, San Diego, 9500 Gilman Drive, La Jolla, CA 92093-0112 USA; e-mail: pfitzsim@ucsd.edu

All other communications to the editors, should be addressed to the Managing Editor, ALEJANDRO ADEM, Department of Mathematics, The University of British Columbia, Room 121, 1984 Mathematics Road, Vancouver, B.C., Canada V6T 1Z2; e-mail: adem@math.ubc.ca

SELECTED PUBLISHED TITLES IN THIS SERIES

1224 **Werner Hoffmann and Satoshi Wakatsuki,** On the Geometric Side of the Arthur Trace Formula for the Symplectic Group of Rank 2, 2018

1223 **Sergey Fomin and Professor Dylan Thurston,** Cluster Algebras and Triangulated Surfaces Part II: Lambda Lengths, 2018

1222 **Maurice Duits and Kurt Johansson,** On Mesoscopic Equilibrium for Linear Statistics in Dyson's Brownian Motion, 2018

1221 **Francesco Lin,** A Morse-Bott Approach to Monopole Floer Homology and the Triangulation Conjecture, 2018

1220 **Paata Ivanisvili, Dmitriy M. Stolyarov, Vasily I. Vasyunin, and Pavel B. Zatitskiy,** Bellman Function for Extremal Problems in BMO II: Evolution, 2018

1219 **Olivier Frécon,** Algebraic $\overline{\mathbb{Q}}$-Groups as Abstract Groups, 2018

1218 **Gabriella Pinzari,** Perihelia Reduction and Global Kolmogorov Tori in the Planetary Problem, 2018

1217 **Chin-Yu Hsiao,** Szegő Kernel Asymptotics for High Power of CR Line Bundles and Kodaira Embedding Theorems on CR Manifolds, 2018

1216 **Robert Lipshitz, Peter Ozsváth, and Dylan P. Thurston,** Bordered Heegaard Floer Homology, 2018

1215 **Lior Fishman, David Simmons, and Mariusz Urbański,** Diophantine Approximation and the Geometry of Limit Sets in Gromov Hyperbolic Metric Spaces, 2018

1214 **David Chataur, Martintxo Saralegi-Aranguren, and Daniel Tanré,** Intersection Cohomology, Simplicial Blow-Up and Rational Homotopy, 2018

1213 **Sebastiano Carpi, Yasuyuki Kawahigashi, Roberto Longo, and Mihály Weiner,** From Vertex Operator Algebras to Conformal Nets and Back, 2018

1212 **Roelof Bruggeman, Youngju Choie, and Nikolaos Diamantis,** Holomorphic Automorphic Forms and Cohomology, 2018

1211 **Shouhei Honda,** Elliptic PDEs on Compact Ricci Limit Spaces and Applications, 2018

1210 **Zhou Gang, Dan Knopf, and Israel Michael Sigal,** Neckpinch Dynamics for Asymmetric Surfaces Evolving by Mean Curvature Flow, 2018

1209 **Cristian Anghel, Iustin Coandă, and Nicolae Manolache,** Globally Generated Vector Bundles with Small c_1 on Projective Spaces, 2018

1208 **Matthew Harrison-Trainor,** Degree Spectra of Relations on a Cone, 2018

1207 **Alastair J. Litterick,** On Non-Generic Finite Subgroups of Exceptional Algebraic Groups, 2018

1206 **Anne-Laure Dalibard and Laure Saint-Raymond,** Mathematical Study of Degenerate Boundary Layers: A Large Scale Ocean Circulation Problem, 2018

1205 **Charles Collot,** Type II Blow Up Manifolds for the Energy Supercritical Semilinear Wave Equation, 2018

1204 **Rui Palma,** Crossed Products by Hecke Pairs, 2018

1203 **Xiao Xiong, Quanhua Xu, and Zhi Yin,** Sobolev, Besov and Triebel-Lizorkin Spaces on Quantum Tori, 2018

1202 **Naiara V. de Paulo and Pedro A. S. Salomão,** Systems of Transversal Sections Near Critical Energy Levels of Hamiltonian Systems in $\mathbb{R}^4$, 2018

1201 **Bernhelm Booß-Bavnbek and Chaofeng Zhu,** The Maslov Index in Symplectic Banach Spaces, 2018

1200 **Francis Nier,** Boundary Conditions and Subelliptic Estimates for Geometric Kramers-Fokker-Planck Operators on Manifolds with Boundaries, 2018

For a complete list of titles in this series, visit the
AMS Bookstore at **www.ams.org/bookstore/memoseries/**.